Projective Group Structures as Absolute Galois Structures with Block Approximation

Number 884

Projective Group Structures as Absolute Galois Structures with Block Approximation

Dan Haran
Moshe Jarden
Florian Pop

September 2007 • Volume 189 • Number 884 (first of 4 numbers) • ISSN 0065-9266

American Mathematical Society
Providence, Rhode Island

2000 *Mathematics Subject Classification.*
Primary 12E30.

Library of Congress Cataloging-in-Publication Data

Haran, Dan.

Projective group structures as absolute Galois structures with block approximation / Dan Haran, Moshe Jarden, Florian Pop.

p. cm. — (Memoirs of the American Mathematical Society, ISSN 0065-9266 ; no. 884)

"September 2007, volume 189, number 884 (first of 4 numbers)."

Includes bibliographical references.

ISBN 978-0-8218-3995-9 (alk. paper)

1. Number theory. 2. Galois theory. 3. Group theory. 4. Polynomials. 5. Field theory (Physics). I. Title.

QA241.H27 2007
512.7—dc22 2007060804

Memoirs of the American Mathematical Society

This journal is devoted entirely to research in pure and applied mathematics.

Subscription information. The 2007 subscription begins with volume 185 and consists of six mailings, each containing one or more numbers. Subscription prices for 2007 are US$649 list, US$519 institutional member. A late charge of 10% of the subscription price will be imposed on orders received from nonmembers after January 1 of the subscription year. Subscribers outside the United States and India must pay a postage surcharge of US$38; subscribers in India must pay a postage surcharge of US$43. Expedited delivery to destinations in North America US$53; elsewhere US$130. Each number may be ordered separately; *please specify number* when ordering an individual number. For prices and titles of recently released numbers, see the New Publications sections of the *Notices of the American Mathematical Society.*

Back number information. For back issues see the *AMS Catalog of Publications.*

Subscriptions and orders should be addressed to the American Mathematical Society, P. O. Box 845904, Boston, MA 02284-5904, USA. *All orders must be accompanied by payment.* Other correspondence should be addressed to 201 Charles Street, Providence, RI 02904-2294, USA.

Memoirs of the American Mathematical Society is published bimonthly (each volume consisting usually of more than one number) by the American Mathematical Society at 201 Charles Street, Providence, RI 02904-2294, USA. Periodicals postage paid at Providence, RI. Postmaster: Send address changes to Memoirs, American Mathematical Society, 201 Charles Street, Providence, RI 02904-2294, USA.

This publication is indexed in *Science Citation Index*®, *SciSearch*®, *Research Alert*®, *CompuMath Citation Index*®, *Current Contents*®*/Physical, Chemical & Earth Sciences.*
Printed in the United States of America.

∞ The paper used in this book is acid-free and falls within the guidelines established to ensure permanence and durability.
Visit the AMS home page at http://www.ams.org/

10 9 8 7 6 5 4 3 2 1 12 11 10 09 08 07

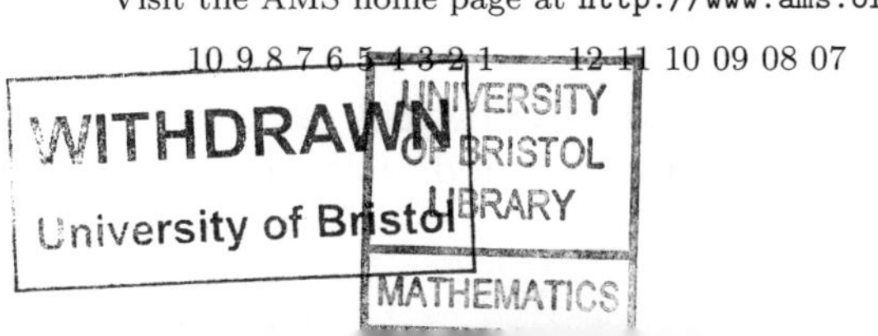

Contents

Abstract

We prove: A proper profinite group structure $\mathbf{G}$ is projective if and only if $\mathbf{G}$ is the absolute Galois group structure of a proper field-valuation structure with block approximation.

Received by the Editor October 12, 2002.

2000 *Mathematics Subject Classification.* 12E30.

Research supported by the Minkowski Center for Geometry at Tel Aviv University, established by the Minerva Foundation.

Introduction

A. Background and Motivation. One of the main features of Field Arithmetic is the interplay between the arithmetic-geometrical properties of a field and the profinite group theoretic properties of its absolute Galois group. Here is the prototype for this kind of results:

BASIC THEOREM.
(a) *If a field K is PAC, then* $\mathrm{Gal}(K)$ *is projective (Ax [FrJ, Thm. 11.6.2]).*
(b) *For every projective group G there exists a field K with* $\mathrm{Gal}(K) \cong G$ *(Lubotzky-v.d.Dries [FrJ, Cor. 23.1.2]).*

Here we say that a field K is **PAC** if every absolutely irreducible variety V defined over K has a K-rational point. By an **absolutely irreducible variety** defined over K we mean a geometrically integral scheme of finite type over K. We denote the separable closure of K by K_s and its algebraic closure by $\tilde{K}$. Then we call $\mathrm{Gal}(K) = \mathrm{Gal}(K_s/K)$ the **absolute Galois group** of K.

A profinite group G is **projective** if every finite embedding problem

$$(1) \qquad (\varphi: G \to A,\ \alpha: B \to A)$$

for G is solvable. Here A and B are finite groups, φ is a homomorphism, and α is an epimorphism. A **solution** of (1) is a homomorphism $\gamma: G \to B$ with $\alpha \circ \gamma = \varphi$.

Both concepts "projective group" and "PAC field" have relative counterparts which we now describe.

Let G be a profinite group and $\mathcal{G}$ a collection of closed subgroups of G. Call G **$\mathcal{G}$-projective** if every finite embedding problem (1) for G has a solution provided for each $\Gamma \in \mathcal{G}$ there is a homomorphism $\gamma_\Gamma: \Gamma \to B$ with $\alpha \circ \gamma_\Gamma = \varphi|_\Gamma$.

Let K be a field and $\mathcal{K}$ a collection of separable algebraic extensions of K. Call K P$\mathcal{K}$C (**pseudo $\mathcal{K}$-closed**) if every smooth absolutely irreducible variety V over K with a K'-rational point for each $K' \in \mathcal{K}$ has a K-rational point.

Both definitions involve local-global principles. Thus, G is $\mathcal{G}$-projective if the existence of local solutions of embedding problems guaranties the existence of global solutions. Analogously, K is P$\mathcal{K}$C if the existence of local points on smooth absolutely irreducible varieties gives global points on them.

It is desirable to generalize the Basic Theorem to the relative case:

TARGET.
(a) *Let K be a field and $\mathcal{K}$ a collection of separable algebraic extensions of K. Put $\mathcal{G} = \{\mathrm{Gal}(K') \mid K' \in \mathcal{K}\}$. Suppose K is P$\mathcal{K}$C. Then* $\mathrm{Gal}(K)$ *is $\mathcal{G}$-projective.*
(b) *Let G be a profinite group and $\mathcal{G}$ a collection of closed subgroups of G. Suppose G is $\mathcal{G}$-projective and for each $\Gamma \in \mathcal{G}$ there exists a field F_Γ with* $\mathrm{Gal}(F_\Gamma) \cong \Gamma$. *Then there exists a field K and an isomorphism $\varphi: G \to$* $\mathrm{Gal}(K)$. *Moreover, for each $\Gamma \in \mathcal{G}$ let K_Γ be the fixed field of $\varphi(\Gamma)$ in K_s. Put $\mathcal{K} = \{K_\Gamma \mid \Gamma \in \mathcal{G}\}$. Then K is P$\mathcal{K}$C.*

The Basic Theorem is a special case of the Target in which both $\mathcal{K}$ and $\mathcal{G}$ are empty.

Another special case of the Target occurs when $\mathcal{K}$ is the collection of all real closures of K and $\mathcal{G}$ is the collection of all subgroups of G which are isomorphic to $\mathrm{Gal}(\mathbb{R})$ [HaJ1, p. 450, Thm.]. In this case P$\mathcal{K}$C fields are referred to as **PRC**

fields. However, in order for Part (b) of the Target to hold, we must assume 1 does not lie in the closure of $\mathcal{G}$; that is, G has an open subgroup U which contains no $\Gamma \in \mathcal{G}$.

Similarly, the Target is reached when $\mathcal{K}$ is the collection of all p-adic closures of K for some fixed prime number p and $\mathcal{G}$ is the collection of all subgroups of G which are isomorphic to $\mathrm{Gal}(\mathbb{Q}_p)$ [HaJ2, p. 148, Thm.]. Again, we must assume 1 does not belong to the closure of $\mathcal{G}$. Then P$\mathcal{K}$C fields are just **PpC fields**.

Another instance where the Target is obtained is when $\mathcal{K} = \{K_1, \dots, K_n\}$ and each K_i is Henselian with respect to a valuation v_i such that $v_1|_K, \dots, v_n|_K$ are independent ([Koe, Thm. 2'] or [HaJ3, Theorems A and B]). Here one starts in Part (b) with a profinite group G which is projective with respect to n closed subgroups $G_1, \dots, G_n$, each of which is isomorphic to the absolute Galois group of a field. Then one constructs K and φ such that the fixed field K_i of $\varphi(G_i)$ is Henselian with respect to a valuation v_i, $i = 1, \dots, n$. Moreover, the restrictions of $v_1, \dots, v_n$ to K are independent.

In general it is possible to prove Part (a) of the Target under some mild compactness assumption on $\mathcal{K}$ [Pop, Thm. 3.3]. We are therefore allowed to make the same assumption on $\mathcal{G}$ in Part (b) of the Target. Nevertheless, when we try to realize G as an absolute Galois group, we are forced to solve certain infinite embedding problems and not only finite ones. So, we must assume G is "strongly $\mathcal{G}$-projective" rather than only $\mathcal{G}$-projective. This has actually been done in [Pop, Thm. 3.4] (Note however that the adjective "strongly" is mistakenly omitted in the formulation of [Pop, Thm. 3.4]). But replacing "G is $\mathcal{G}$-projective" by "G is strongly $\mathcal{G}$-projective" in Part (b) brings the Target out of balance. To restore the balance we allow adding extra conditions to (a) and to (b). The rule is that each assumption we make on $\mathcal{K}$ in (a) should appear as a consequence in (b). Similarly, each assumption we make on $\mathcal{G}$ in (b) should appear as a consequence in (a). The disturbed balance in [Pop] is restored only when "large quotients" exist, as in the case of p-adically closed fields [Pop, Section 1, Lemma and Definition]. The general case is left unbalanced in [Pop].

The goal of this work is to achieve a very general balanced Target. Like in the above mentioned three instances, we extend both $\mathcal{K}$ and $\mathcal{G}$ to "structures" over a profinite space X and let each field in $\mathcal{K}$ be a Henselian closure of a valuation of the base field K. In order to prove projectivity of the group structure in (a) we must assume a strong form of the weak approximation theorem. We call it the "block approximation condition". One of the main achievements of this work is the realization of the structure in (b) as an "absolute Galois structure"of a "field-valuation structure" satisfying the block approximation condition.

B. The main theorem. For the convenience of the reader we state the main result of this work, define all concepts appearing in it, and describe the most essential ingredients of the proof.

MAIN THEOREM.

(a) *Let* $\mathbf{K} = (K, X, K_x, v_x)_{x \in X}$ *be a proper Henselian field-valuation structure. Suppose* $\mathbf{K}$ *satisfies the block approximation condition. Then* $\mathrm{Gal}(\mathbf{K}) = (\mathrm{Gal}(K), X, \mathrm{Gal}(K_x))_{x \in X}$ *is a proper projective group structure.*

(b) *Let* $\mathbf{G} = (G, X, G_x)_{x \in X}$ *be a proper projective group structure and* $\bar{\kappa}\colon \mathbf{G} \to \mathrm{Gal}(\bar{\mathbf{K}})$ *be a Galois approximation of* $\mathbf{G}$*. Then there exists a proper Henselian field-valuation structure* $\mathbf{K} = (K, X, K_x, v_x)_{x \in X}$ *satisfying the block* *approximation condition and there is an isomorphism* $\kappa: \mathbf{G} \to \mathrm{Gal}(\mathbf{K})$ *such that* $\mathrm{res} \circ \kappa = \bar{\kappa}$.

Here are the definitions of the notions which occur in the Main Theorem.

We call $\mathbf{G} = (G, X, G_x)_{x \in X}$ a **group structure** if G is a profinite group, X is a profinite space, and for each $x \in X$, G_x is a closed subgroup of G satisfying these conditions:

(2a) G acts continuously on X from the right.

(2b) $G_{x^g} = G_x^g$ for all $x \in X$ and $g \in G$.

(2c) Let $\mathrm{Subgr}(G)$ be the space of all closed subgroups of G equipped with the **étale topology**. (A basis of the étale topology consists of all sets $\mathrm{Subgr}(U)$ with U open in G.) Then the map $\delta_{\mathbf{G}}: X \to \mathrm{Subgr}(G)$ defined by $\delta_{\mathbf{G}}(x) = G_x$ is continuous in the étale topology.

(2d) $\{g \in G \mid x^g = x\} \leq G_x$ for each $x \in X$.

We say $\mathbf{G}$ is **proper** if the map $\delta_{\mathbf{G}}: X \to \{G_x \mid x \in X\}$ is a homeomorphism in the étale topology.

A group structure $\mathbf{G}$ is **projective** if every finite embedding problem

$$(\varphi: \mathbf{G} \to \mathbf{A},\ \alpha: \mathbf{B} \to \mathbf{A}) \tag{3}$$

for $\mathbf{G}$ is solvable. Here we call (3) an **embedding problem** if the following holds:

(4a) $\mathbf{A} = (A, I, A_i)_{i \in I}$ and $\mathbf{B} = (B, J, B_j)_{j \in J}$ are **finite group structures**, i.e. A, B, I, and J are finite.

(4b) $\varphi: \mathbf{G} \to \mathbf{A}$ is a **morphism**; that is, φ is a pair consisting of a homomorphism $\varphi: G \to A$ and a continuous map $\varphi: X \to I$ such that $\varphi(x^g) = \varphi(x)^{\varphi(g)}$ and $\varphi(G_x) \leq A_{\varphi(x)}$ for all $x \in X$ and $g \in G$.

(4c) $\alpha: \mathbf{B} \to \mathbf{A}$ is a **cover**; that is, α is a morphism, $\alpha(B) = A$, $\alpha(J) = I$, $\alpha: B_j \to A_{\alpha(j)}$ is an isomorphism for each $j \in J$, and for all $j_1, j_2 \in J$ with $\varphi(j_1) = \varphi(j_2)$ there is $b \in \mathrm{Ker}(\alpha)$ with $j_1^b = j_2$.

A **solution** of (3) is a morphism $\gamma: \mathbf{G} \to \mathbf{B}$ satisfying $\alpha \circ \gamma = \varphi$.

We call $(K, X, K_x)_{x \in X}$ a **field structure** if K is a field, X is a profinite space, and K_x is a separable algebraic extension, $x \in X$, such that

$$\mathrm{Gal}(\mathbf{K}) = (\mathrm{Gal}(K), X, \mathrm{Gal}(K_x))_{x \in X}$$

is a group structure.

A **Galois approximation** of a group structure $\mathbf{G} = (G, X, G_x)_{x \in X}$ is a morphism $\bar{\kappa}: \mathbf{G} \to \mathrm{Gal}(\bar{\mathbf{K}})$ where $\bar{\mathbf{K}} = (\bar{K}, \bar{X}, \bar{K}_{\bar{x}})_{\bar{x} \in \bar{X}}$ is a field structure, $\bar{\kappa}(G) = \mathrm{Gal}(\bar{K})$, $\bar{\kappa}(X) = \bar{X}$, and $\bar{\kappa}: G_x \to \mathrm{Gal}(\bar{K}_{\bar{\kappa}(x)})$ is an isomorphism for each $x \in X$.

We call $\mathbf{K} = (K, X, K_x, v_x)_{x \in X}$ a **field-valuation structure** if $(K, X, K_x)_{x \in X}$ is a field structure and v_x is a valuation of K_x satisfying these conditions:

(5a) $v_{x^\sigma} = v_x^\sigma$ for all $x \in X$ and $\sigma \in \mathrm{Gal}(K)$.

(5b) For each finite separable extension L the map $\nu_L: X_L \to \mathrm{Val}(L)$ given by $\nu_L(x) = v_x|_L$ is continuous. Here $X_L = \{x \in X \mid L \subseteq K_x\}$ and $\mathrm{Val}(L)$ is the space of all valuation of L including the trivial one. A subbasis for the topology of $\mathrm{Val}(L)$ is the collection of all sets

$$U = \{w \in \mathrm{Val}(L) \mid w(a) > 0\} \quad \text{and} \quad U' = \{w \in \mathrm{Val}(L) \mid w(a) \geq 0\}$$

with $a \in L$.

We say that $\mathbf{K}$ is **Henselian**, if in addition (K_x, v_x) is **Henselian** for each $x \in X$.

A **block approximation problem** for $\mathbf{K}$ is a data $(V, X_i, L_i, \mathbf{a}_i, c_i)_{i \in I_0}$ satisfying these conditions:

(6a) I_0 is a finite set.
(6b) X_i is an open-closed subset of X, $i \in I_0$.
(6c) L_i is a finite separable extension of K contained in K_x for all $x \in X_i$ and $i \in I_0$.
(6d) $\mathrm{Gal}(L_i) = \{\sigma \in \mathrm{Gal}(L) \mid X_i^\sigma = X_i\}$, $i \in I_0$.
(6e) For each $i \in I_0$ let R_i be a subset of $\mathrm{Gal}(K)$ satisfying $\mathrm{Gal}(K) = \dot\bigcup_{\rho \in R_i} \mathrm{Gal}(L_i)\rho$. Then $X = \dot\bigcup_{i \in I_0} \dot\bigcup_{\rho \in R_i} X_i^\rho$.
(6f) V is a smooth affine absolutely irreducible variety over K.
(6g) $\mathbf{a}_i \in V(L_i)$, $i \in I_0$.
(6h) $c_i \in K^\times$, $i \in I_0$.

A **solution** of the block approximation problem is a point $\mathbf{a} \in V(K)$ satisfying $v_x(\mathbf{a} - \mathbf{a}_i) > v(c_i)$ for all $i \in I_0$ and $x \in X_i$. We say that $\mathbf{K}$ satisfies the **block approximation condition** if every block approximation problem for $\mathbf{K}$ has a solution.

Finally, in the notation of (b) of the Main Theorem, we say that κ **lifts** $\bar\kappa$ if K is a regular extension of $\bar K$ and the epimorphism $\mathrm{res}\colon \mathrm{Gal}(K) \to \mathrm{Gal}(\bar K)$ extends to a morphism $\rho\colon \mathrm{Gal}(\mathbf{K}) \to \mathrm{Gal}(\bar{\mathbf{K}})$ with $\rho \circ \kappa = \bar\kappa$.

In the rest of the introduction we explain some of the main points of the proof. This will partially explain why the notions in the Main Theorem are so involved.

In the proof of Part (b) of the Target we have to solve embedding problems of the type $(\varphi\colon G \to \mathrm{Gal}(K), \alpha\colon \mathrm{Gal}(L) \to \mathrm{Gal}(K))$. Since $\mathrm{Gal}(K)$ and $\mathrm{Gal}(L)$ are infinite, it does not follow immediately from the projectivity of G that a solution γ exists. However, a result of Gruenberg [FrJ, Lemma 22.3.2] does give γ in the setup of the Basic Theorem. In all other cases of the Target Theorem proved prior to this work it is needed that for each $\Gamma \in \mathcal{G}$, $\gamma(\Gamma)$ belongs to a subset of $\mathrm{Subgr}(G)$ given in advance. Therefore, the profinite groups G have been equipped with certain group structures and homomorphisms have been replaced by morphisms such that solvability of finite embedding problems in the so obtained category implies solvability of arbitrary embedding problems.

Each of these structures consisted of a profinite group G acting on a profinite space and local objects parametrized by X. It was further assumed that the action of G on X is regular; that is $x^g = x$ for $x \in X$ and $g \in G$ implies $g = 1$. This gave a closed system of representatives for the G-orbits of X [HaJ2, Lemma 2.4]. But in general, closed system of representatives do not exist. Instead we find representatives modulo each open normal subgroup of G. More precisely, let $\mathbf{G} = (G, X, G_x)_{x \in X}$ be a group structure as in the Main Theorem and N an open normal subgroup of G. Then we find a finite system of triples $(G_i, X_i, R_i)_{i \in I_0}$ which we call a **special partition** of G. It satisfies the following conditions:

(7a) I_0 is a finite set, disjoint from X.
(7b) X_i is an open-closed subset of X, $i \in I_0$.
(7c) G_i is an open subgroup of G containing G_x for all $x \in X_i$, $i \in I_0$.
(7d) $G_i = \{\sigma \in \mathrm{Gal}(L) \mid X_i^\sigma = X_i\}$, $i \in I_0$.
(7e) R_i is finite, $G = \dot\bigcup_{\rho \in R_i} G_i\rho$, and $X = \dot\bigcup_{i \in I_0} \dot\bigcup_{\rho \in R_i} X_i^\rho$.

The existence of special partitions goes back to [Pop, Prop. 4.9].

We use special partitions on several occasions:

(8a) to extend each homomorphism $\varphi\colon G \to A$ with a finite group A to a morphism $\varphi\colon \mathbf{G} \to \mathbf{A}$ where $\mathbf{A} = (A, I, A_i)_{i\in I}$ is a finite group structure given in advance (Lemma 3.7);

(8b) in the definition of "unirational arithmetical problem" (Section 6) and "block approximation problem" (Section 12) and in the proof of Part (a) of the Main Theorem (Lemma 14.2); and

(8c) in the proof of Part (b) of the Main Theorem (Lemma 15.1).

A second essential ingredient in the proof of Part (a) of the Main Theorem is the local homeomorphism theorem for étale morphisms of varieties over Henselian fields [GPR, Thm. 9.4]. A special partition, a "locally uniform Hensel's lemma" (Corollary 10.4), and block approximation prepare the use of the local homeomorphism theorem. The idea to use this set up goes back to [HaJ3, Prop. 3.2]. Block approximation can be found in [FHV, Prop. 2.1] in the context of real closed fields.

In a subsequent work we intend to apply the Main Theorem to prove the Target Theorem in a general p-adic setting which will make a far reaching generalization of [HaJ1].

1. Étale Topology

Let G be a profinite group. Denote the collection of all closed (resp. open, open normal) subgroups of G by $\mathrm{Subgr}(G)$ (resp. $\mathrm{Open}(G)$, $\mathrm{OpenNormal}(G)$). We introduce two topologies on $\mathrm{Subgr}(G)$, the strict topology and the étale topology, and relate them to each other.

A basis of the **strict topology** is the collection of all sets

$$\nu(H, N) = \{A \in \mathrm{Subgr}(G) \mid AN = HN\}, \tag{1}$$

with $H \in \mathrm{Open}(G)$ and $N \in \mathrm{OpenNormal}(G)$. When G is finite, the strict topology is the discrete topology. In general, $\mathrm{Subgr}(G) \cong \varprojlim \mathrm{Subgr}(G/N)$ with N ranging over all open normal subgroups of G. Thus, $\mathrm{Subgr}(G)$ is a profinite space under the strict topology. Indeed, each of the sets $\nu(H, N)$ is also closed in the strict topology. We use the adverb "strictly" as a substitute for "in the strict topology". For example, given a subset $\mathcal{G}$ of $\mathrm{Subgr}(G)$, we say $\mathcal{G}$ is **strictly open** (resp. **closed**, **compact**, **Hausdorff**) if it is open (resp. closed, compact, Hausdorff) in the strict topology. Likewise, for a function f from a topological space X into $\mathrm{Subgr}(G)$ we say f is **strictly continuous** if f is continuous when $\mathrm{Subgr}(G)$ is equipped with the strict topology.

A basis of the **étale topology** is the collection of all sets

$$\{\mathrm{Subgr}(U) \mid U \in \mathrm{Open}(G)\}$$

with $U \in \mathrm{Open}(G)$. As above, for a subset $\mathcal{G}$ of $\mathrm{Subgr}(G)$ we say $\mathcal{G}$ is **étale open** (**closed**, **compact**, **Hausdorff**, etc) if $\mathcal{G}$ is open (closed, compact, Hausdorff, etc) in the étale topology. Likewise, for a function f from a topological space X into $\mathrm{Subgr}(G)$ we say f is **étale continuous** if f is continuous when $\mathrm{Subgr}(G)$ is equipped with the étale topology.

Note: We use the adjective **compact** for a topological space X in the sense of Hewitt-Ross [HRo]. Thus, every open covering of X has a finite subcovering (but, in contrast to the terminology of Bourbaki, X need not be Hausdorff).

REMARK 1.1. *Categorical properties of the étale topology.*

(a) Subgroups: Let H be a closed subgroup of G. Then a subgroup H_0 of H is open in H if and only if $H_0 = H \cap G_0$ with $G_0 \in \mathrm{Open}(G)$. Moreover, $\mathrm{Subgr}(H_0) = \mathrm{Subgr}(H) \cap \mathrm{Subgr}(G_0)$. Thus, the étale topology of $\mathrm{Subgr}(H)$ is the one induced from the étale topology of $\mathrm{Subgr}(G)$.

(b) Quotients: Let N be a closed normal subgroup of G. Put $\bar{G} = G/N$ and let $\pi\colon G \to \bar{G}$ be the quotient map. Given $\bar{U} \in \mathrm{Open}(\bar{G})$, put $U = \pi^{-1}(\bar{U})$ and observe that $\pi^{-1}(\mathrm{Subgr}(\bar{U})) = \mathrm{Subgr}(U)$. It follows that the étale topology of $\mathrm{Subgr}(\bar{G})$ is the quotient topology of $\mathrm{Subgr}(G)$ via the quotient map $\pi\colon \mathrm{Subgr}(G) \to \mathrm{Subgr}(\bar{G})$. $\square$

REMARK 1.2. *Étale versus strict.* The strict topology of $\mathrm{Subgr}(G)$ is finer than the étale topology. Indeed, consider an open subgroup U of G. Choose an open normal subgroup N of G in U. List the subgroups between N and U as $H_1, \ldots, H_n$. Then $\mathrm{Subgr}(U) = \bigcup_{i=1}^n \{A \in \mathrm{Subgr}(G) \mid AN = H_i\}$. Hence, $\mathrm{Subgr}(U)$ is strictly open (and closed).

Since $\mathrm{Subgr}(G)$ is strictly profinite, this gives the following chain of implications for a subset $\mathcal{G}$ of $\mathrm{Subgr}(G)$: $\mathcal{G}$ is étale closed $\Longrightarrow$ $\mathcal{G}$ is strictly closed $\Longleftrightarrow$ $\mathcal{G}$ is strictly compact $\Longrightarrow$ $\mathcal{G}$ is étale compact. $\square$

The intersection of two étale open basic sets contains the trivial group. So, if $G \neq 1$, the étale topology of $\mathrm{Subgr}(G)$ is not Hausdorff. However, a subset $\mathcal{G}$ of $\mathrm{Subgr}(G)$ can be étale Hausdorff. Indeed, we will be looking for such $\mathcal{G}$ which are even étale profinite.

Denote the strict closure of a subset $\mathcal{G}$ of $\mathrm{Subgr}(G)$ (resp. a point $H \in \mathrm{Subgr}(G)$) by $\mathrm{StrictClosure}(\mathcal{G})$ (resp. $\mathrm{StrictClosure}(H)$).

LEMMA 1.3. *Let $\mathcal{G}$ be a subset of* $\mathrm{Subgr}(G)$.

(a) *Let $H, H' \in \mathcal{G}$. Suppose $H \cap H'$ contains no L which belongs to* $\mathrm{StrictClosure}(\mathcal{G})$. *Then H and H' can be separated by the étale topology of $\mathcal{G}$.*

(b) *Suppose $H \cap H'$ contains no $L \in$* $\mathrm{StrictClosure}(\mathcal{G})$ *for all distinct $H, H' \in \mathcal{G}$. Then $\mathcal{G}$ is étale Hausdorff.*

PROOF. Statement (b) follows from (a). So, we prove (a). Assume H and H' cannot be separated by the étale topology of $\mathcal{G}$. Denote the set of all pairs $(U, U') \in \mathrm{Open}(G) \times \mathrm{Open}(G)$ with $H \leq U$ and $H' \leq U'$ by $\mathcal{U}$. Then, $\mathrm{Subgr}(U) \cap \mathrm{Subgr}(U') \cap \mathcal{G} \neq \emptyset$ for all $(U, U') \in \mathcal{U}$. Hence, $\mathrm{Subgr}(U) \cap \mathrm{Subgr}(U') \cap \mathrm{StrictClosure}(\mathcal{G}) \neq \emptyset$ for all $(U, U') \in \mathcal{U}$. Each of the sets $\mathrm{Subgr}(U) \cap \mathrm{Subgr}(U') \cap \mathrm{StrictClosure}(\mathcal{G})$ is strictly closed (Remark 1.2). The intersection of finitely many of them is a set of the same type. Hence, the intersection is nonempty. Since $\mathrm{StrictClosure}(\mathcal{G})$ is strictly compact, there is an $L \in \bigcap_{(U,U') \in \mathcal{U}} \mathrm{Subgr}(U) \cap \mathrm{Subgr}(U') \cap \mathrm{StrictClosure}(\mathcal{G})$. It satisfies $L \leq H \cap H'$. This contradicts the assumption of the lemma. $\square$

COROLLARY 1.4. *Let $\mathcal{G}$ be a subset of* $\mathrm{Subgr}(G)$ *with* $1 \notin \mathrm{StrictClosure}(\mathcal{G})$.

(a) *Let $H, H' \in \mathcal{G}$. Suppose $H \cap H' = 1$. Then H and H' can be separated by the étale topology of $\mathcal{G}$.*

(b) *Suppose $H \cap H' = 1$ for all distinct $H, H' \in \mathcal{G}$. Then $\mathcal{G}$ is étale Hausdorff.*

Here is a certain converse to Corollary 1.4:

LEMMA 1.5. *Let G be a profinite group and $\mathcal{G}$ a subset of* $\mathrm{Subgr}(G)$. *Suppose $\mathcal{G}$ is étale Hausdorff and contains at least two groups. Then* $1 \notin \mathrm{StrictClosure}(\mathcal{G})$.

PROOF. Let H_1 and H_2 be distinct groups in $\mathcal{G}$. Then there are disjoint étale open subsets $\mathcal{U}_1$ and $\mathcal{U}_2$ of $\mathcal{G}$ such that $H_i \in \mathcal{U}_i$, $i = 1, 2$. For each i there is a $U_i \in \mathrm{Open}(G)$ with $H_i \in \mathcal{G} \cap \mathrm{Subgr}(U_i) \subseteq \mathcal{U}_i$. Let $U = U_1 \cap U_2$. Then $U \in \mathrm{Open}(G)$ and

$$\mathcal{G} \cap \mathrm{Subgr}(U) \subseteq \mathcal{G} \cap \mathrm{Subgr}(U_1) \cap \mathrm{Subgr}(U_2) \subseteq \mathcal{U}_1 \cap \mathcal{U}_2 = \emptyset.$$

It follows, $1 \notin \mathrm{StrictClosure}(\mathcal{G})$. □

2. Group Structures

The profinite group structures we introduce in this section replace the Artin-Schreier Structures of [HaJ1], the Γ-structures of [HaJ2], and the étale spaces of [Har]. The category of profinite group structures admits quotients (Example 2.5), fiber products (Construction 2.9), and inverse limits (Remark 2.7). These are the necessary tools to prove that solvability of finite embedding problems of a finite group structure $\mathbf{G}$ implies that the solvability of arbitrary embedding problems for $\mathbf{G}$ (Proposition 4.2).

A **topological group space** (also called a **group space**) is a pair (X, G) consisting of a topological space X, a topological group G, and a continuous action of G on X from the right (which we write exponentially). If X is a profinite space and G is a profinite group, we say (X, G) is a **profinite group space**. A **morphism** $\varphi\colon (X, G) \to (Y, H)$ of group spaces is a couple consisting of a continuous map $\varphi\colon X \to Y$ and a continuous group homomorphism $\varphi\colon G \to H$ satisfying $\varphi(x^g) = \varphi(x)^{\varphi(g)}$ for all $x \in X$ and $g \in G$. Composition of morphisms of group spaces and the identity maps are morphisms of profinite group spaces satisfying the associativity law. Thus, the class of topological (resp. profinite) groups spaces with their morphisms form a category.

For each group space (X, G) and each element $x \in X$, we let $S_x = \{g \in X \mid x^g = x\}$. It is a closed subgroup of G called the **stabilizer** of x. For each $\sigma \in G$ we have $S_{x^\sigma} = S_x^\sigma$. If $\varphi\colon (X, G) \to (Y, H)$ is a morphism and $x \in X$, then $\varphi(S_x) \le S_{\varphi(x)}$.

Every profinite group G acts on $\mathrm{Subgr}(G)$ by conjugation. This action is strictly continuous as well as étale continuous. Therefore, $(\mathrm{Subgr}(G), G)$ with $\mathrm{Subgr}(G)$ equipped with the strict topology (resp. the étale topology) is a profinite (resp. topological) group space. In Section 6 we encounter our second basic example of profinite group spaces arising in the context of absolute Galois groups.

A **profinite group structure** is a triple $\mathbf{G} = (G, X, \delta)$ consisting of a profinite group space (X, G) and an étale continuous map $\delta\colon X \to \mathrm{Subgr}(G)$. This object must satisfy the following conditions:

(1a) $G_{x^g} = G_x^g$ for all $x \in X$ and $g \in G$; thus δ is a morphism of group spaces.

(1b) $S_x \le G_x$ for each $x \in X$.

Denote δ also by $\delta_{\mathbf{G}}$. The continuity of $\delta_{\mathbf{G}}$ means that $\{x \in X \mid G_x \le U\}$ is an open subset of X for each $U \in \mathrm{Open}(G)$.

We write $\mathbf{G}$ also as $(G, X, G_x)_{x \in X}$ and refer to $\mathbf{G}$ as a **group structure** (omitting "profinite").

A **morphism** of group structures

$$\varphi\colon (G, X, G_x)_{x \in X} \to (H, Y, H_y)_{y \in Y} \tag{2}$$

is a morphism $\varphi\colon (X, G) \to (Y, H)$ of profinite group spaces such that $\varphi(G_x) \le H_{\varphi(x)}$.

We call φ an **epimorphism** if $\varphi(G) = H$, $\varphi(X) = Y$, and for each $y \in Y$ there is $x \in X$ with $\varphi(x) = y$ and $\varphi(G_x) = H_y$.

We call φ a **cover** if φ is an epimorphism with the following properties:

(3a) φ maps each G_x isomorphically onto $G_{\varphi(x)}$.

(3b) $\varphi(x) = \varphi(x')$ implies $x^k = x'$ for some $k \in \mathrm{Ker}(\varphi)$.

If an epimorphism $\varphi: \mathbf{G} \to \mathbf{H}$ satisfies (3a) (but not necessarily (3b)), we say φ is **rigid**. We call **G finite**, if both G and X are finite.

REMARK 2.1. *Proper group structures.* Let $\mathbf{G} = (G, X, G_x)_{x\in X}$ be a group structure. Write $\mathcal{G} = \{G_x \mid x \in X\}$. We say **G** is **proper**, if $\delta_{\mathbf{G}}: X \to \mathcal{G}$ is an étale homeomorphism. Then $\mathcal{G}$ is étale profinite. Moreover, $S_x = G_x$ for each $x \in X$. Indeed, if $g \in G_x$, then $G_{x^g} = G_x^g = G_x$, hence $x^g = x$. Thus, $N_G(\Gamma) = \Gamma$ for each $\Gamma \in \mathcal{G}$. If $X = \{x\}$ consists of one element and $\sigma \in G$, then $x^\sigma = x$, hence $\sigma \in S_x = G_x$. Therefore, $G_x = G$. If X contains at least two points, then $1 \notin \mathrm{StrictClosure}(\mathcal{G})$ (Lemma 1.5).

Let $\mathbf{H} = (H, Y, H_y)_{y\in Y}$ be another proper group structure and $\varphi: G \to H$ a group homomorphism. Put $\mathcal{H} = \{H_y \mid y \in Y\}$. Suppose $\varphi(\mathcal{G}) \subseteq \mathcal{H}$. Then $\delta_{\mathbf{H}}^{-1} \circ \varphi \circ \delta_{\mathbf{G}}$ is a continuous map from X into Y which is compatible with the action of G and H. This gives a unique extension of $\varphi: G \to H$ to a morphism $\varphi: \mathbf{G} \to \mathbf{H}$ satisfying $\varphi(G_x) = H_{\varphi(x)}$ for each $x \in X$.

Now consider a third proper group structure $\mathbf{A} = (A, I, A_i)_{i\in I}$. Let $\alpha: \mathbf{G} \to \mathbf{A}$ and $\beta: \mathbf{H} \to \mathbf{A}$ be morphisms. Suppose $\varphi(G_x) = H_{\varphi(x)}$, $\beta(H_y) = A_{\beta(y)}$, and $\alpha(G_x) = A_{\alpha(x)}$ for all $x \in X$ and $y \in Y$. Then $\alpha = \beta \circ \varphi$ as homomorphisms of groups implies $\alpha = \beta \circ \varphi$ as morphisms of group structures.

Finally suppose $\varphi: \mathbf{G} \to \mathbf{H}$ is a rigid epimorphism of group structures with **G** proper and $\mathrm{Ker}(\varphi) = 1$. Then φ is an isomorphism and **H** is proper. Indeed, $\varphi: G \to H$ is an isomorphism. It remains to prove that $\varphi: X \to Y$ is an isomorphism. Since both X and Y are profinite spaces and $\varphi: X \to Y$ is continuous and surjective, it suffices to prove that $\varphi: X \to Y$ is injective. Consider $x, x' \in X$ with $\varphi(x) = \varphi(x')$. Then $\varphi(G_x) = \varphi(G_{x'})$. Hence, $G_x = G_{x'}$. Since **G** is proper, $x = x'$, as desired. □

LEMMA 2.2. *Suppose (2) is a cover of groups structures. Then $\varphi(S_x) = S_{\varphi(x)}$ for each $x \in X$. In particular, if $H_y = S_y$ for each $y \in Y$, then $G_x = S_x$ for each $x \in X$.*

PROOF. Let $x \in X$ and $y = \varphi(x)$. We have already mentioned that $\varphi(S_x) \le S_y$. Also, $\varphi: G_x \to H_y$ is an isomorphism. Hence, in order to prove that $\varphi(S_x) = S_y$, it suffices to consider $g \in \varphi^{-1}(S_y) \cap G_x$ and to prove that $g \in S_x$.

Indeed, $\varphi(x^g) = y^{\varphi(g)} = y = \varphi(x)$. Hence, there is a $k \in \mathrm{Ker}(\varphi)$ with $x^{gk} = x$. Thus, $gk \in S_x$. Therefore, $k \in \mathrm{Ker}(\varphi) \cap G_x = 1$. It follows that $g \in S_x$.

Now assume $H_y = S_y$. Then, by the preceding paragraph, $\varphi(G_x) = H_y = \varphi(S_x)$. Since, $\varphi: G_x \to H_y$ is an isomorphism, $G_x = S_x$, as claimed. □

LEMMA 2.3. *Let $(G, X, G_x)_{x\in X}$ be a group structure and Y a closed subset of X. Then $\bigcup_{x\in Y} G_x$ is closed in G.*

PROOF (After [Gil, Lemma 1.4]). Let $g \in G \smallsetminus \bigcup_{x\in Y} G_x$. For each $x \in Y$ there is an open normal subgroup N_x of G with $gN_x \cap G_x = \emptyset$. Thus, $g \notin G_xN_x$. Since $G_xN_x \in \mathrm{Open}(G)$, continuity of $\delta_{\mathbf{G}}$ implies $V_x = \{y \in Y \mid G_y \le G_xN_x\}$ is an open neighborhood of x in Y. Since Y is compact, the covering $\{V_x \mid x \in Y\}$

has a finite subcovering $\{V_{x_1}, \dots, V_{x_n}\}$. Then $N = \bigcap_{i=1}^n N_{x_i}$ is an open normal subgroup of G and $g \notin G_y N$ for each $y \in Y$. Therefore, $gN \subseteq G \smallsetminus \bigcup_{y\in Y} G_y$. $\square$

Proper group structures are our main subject of research. We have introduced the more general concept of group structures in order to be able to extend the basic operations of the category of profinite groups to the category of group structures. This is not always possible in the category of proper group structures. For example, a quotient of a proper group structure need not be proper (Example 2.5).

EXAMPLE 2.4. *Absolute Galois group structures.* An **absolute Galois group structure** is a group structure $\mathbf{G} = (\mathrm{Gal}(K), X, \mathrm{Gal}(K_x))_{x\in X}$ where each K_x is a separable algebraic extension of K. Let $\mathbf{H} = (\mathrm{Gal}(L), Y, L_y)_{y\in Y}$ be another absolute Galois group structure. Suppose both $\mathbf{G}$ and $\mathbf{H}$ are proper, $K \subseteq L$ and for each $y \in Y$ there is an $x \in X$ with $L_y \cap K_s = K_x$. By Remark 2.1, $\mathrm{res}_{L_s/K_s}\colon \mathrm{Gal}(L) \to \mathrm{Gal}(K)$ extends to a unique morphism $\rho\colon \mathbf{H} \to \mathbf{G}$ of group structures satisfying $\mathrm{res}_{L_s/K_s}(\mathrm{Gal}(L_y)) = \mathrm{Gal}(K_x)$ for all $y \in Y$ and $x = \rho(y)$. We denote this morphism by res if the reference to K and L is clear from the context. $\square$

EXAMPLE 2.5. *Quotient maps.* Let $\mathbf{G} = (G, X, G_x)_{x\in X}$ be a group structure and N a closed normal subgroup of G. Put $\bar{G} = G/N$ and $\bar{X} = X/N$. Let $\pi\colon G \to \bar{G}$ and $\pi\colon X \to \bar{X}$ the quotient maps: $\pi(g) = \bar{g} = gN$ and $\pi(x) = \bar{x} = \{x^\nu \mid \nu \in N\}$. Then $\bar{X}$ is a profinite space [HaJ1, Claim 1.6]. For each $x \in X$ let $\bar{G}_{\bar{x}} = \pi(G_x) = G_xN/N$.

Consider $\bar{U} \in \mathrm{Open}(\bar{G})$. Put $U = \pi^{-1}(\bar{U})$. Then $\pi^{-1}(\{\bar{x} \in \bar{X} \mid \bar{G}_{\bar{x}} \le \bar{U}\}) = \{x \in X \mid G_x \le U\}$. Hence, the map $\delta_{\bar{\mathbf{G}}}\colon \bar{X} \to \mathrm{Subgr}(\bar{G})$ given by $\delta_{\bar{\mathbf{G}}}(\bar{x}) = \bar{G}_{\bar{x}}$ is étale continuous. Also, $\bar{G}$ acts continuously on $\bar{X}$ by $\bar{x}^{\bar{g}} = \bar{x}$ and $\bar{x}^{\bar{\sigma}} = \bar{x}$ implies $\bar{\sigma} \in \bar{G}_{\bar{x}}$. Thus, $\bar{\mathbf{G}} = (\bar{G}, \bar{X}, \bar{G}_{\bar{x}})_{\bar{x}\in\bar{X}}$ is a group structure which we denote by $\mathbf{G}/N$ and $\pi\colon \mathbf{G} \to \bar{\mathbf{G}}$ is an epimorphism. Moreover, $\pi(G_x) = G_{\pi(x)}$ for every $x \in X$. We call π the **quotient map**. If $G_x \cap N = 1$ for each $x \in X$, then π is a cover.

Let $\mathcal{G} = \{G_x \mid x \in X\}$ and $\bar{\mathcal{G}} = \{\bar{G}_{\bar{x}} \mid \bar{x} \in \bar{X}\}$. Then π induces a strictly continuous map of $\mathrm{Subgr}(G)$ onto $\mathrm{Subgr}(\bar{G})$ and $\pi(\mathcal{G}) = \bar{\mathcal{G}}$. Thus, if $1 \notin \mathrm{StrictClosure}(\bar{\mathcal{G}})$, then $1 \notin \mathrm{StrictClosure}(\mathcal{G})$.

Conversely, every cover $\varphi\colon \mathbf{G} \to \mathbf{H}$ of group structures is isomorphic to the quotient map $\mathbf{G} \to \mathbf{G}/\mathrm{Ker}(\varphi)$. Indeed, let $\mathbf{H} = (H, Y, H_y)_{y\in Y}$. Then φ induces a bijective continuous map $\bar{\varphi}\colon \bar{X} \to Y$. Since both $\bar{X}$ and Y are profinite, $\bar{\varphi}$ is a homeomorphism.

Now consider the case where $N = G$. Suppose $|\bar{X}| > 1$. Then $\bar{G} = 1$ and the map $\delta_{\bar{\mathbf{G}}}$ is not injective. Thus, $\bar{\mathbf{G}}$ need not be proper even if G is proper. This is one of the reasons why we work in the category of group structures and not in the category of proper group structures, which may look at first glance more attractive. Another reason is the need to use morphisms called "Galois approximations" (Section 14). The target objects of Galois approximations are group structures which need not be proper.

Quotient maps of group structures has the universal property of quotient maps of groups. Thus, if $\pi\colon \mathbf{G} \to \bar{\mathbf{G}}$ and $\varphi\colon \mathbf{G} \to \mathbf{H}$ are quotient maps satisfying $\mathrm{Ker}(\varphi) \le \mathrm{Ker}(\pi)$, then there is a unique quotient map $\psi\colon \mathbf{H} \to \mathbf{G}/N$ satisfying $\psi \circ \varphi = \pi$. Moreover, π is a cover if and only if φ and ψ are covers. Finally, if $N' \le N$ are closed normal subgroups of G, then there is a natural isomorphism $\mathbf{G}/N \cong (\mathbf{G}/N')/(N/N')$. $\square$

REMARK 2.6. *Sub-group-structures.* Let $\mathbf{G} = (G, X, G_x)_{x\in X}$ and $\mathbf{H} = (H, Y, H_y)_{y\in Y}$ be group structures. We say $\mathbf{H}$ is a **sub-group-structure** of $\mathbf{G}$ if $H \le G$, Y is a subspace of X, and $H_y = G_y$ for each $y \in Y$. If $\mathbf{G}$ proper, then so is $\mathbf{H}$.

Suppose we start with a group G, a profinite space X, and for each $x \in X$ a closed subgroup G_x of G. Consider a closed subgroup H of G which contains each G_x. If U is an open subgroup of G, then $V = U \cap H$ is an open subgroup of H. Conversely, for each open subgroup V of H there is an open subgroup U of G with $V = U \cap H$. In each case $\{x \in X \mid G_x \le U\} = \{x \in X \mid G_x \le V\}$. Hence, if one of the sets is open, so is the other. Thus, $(G, X, G_x)_{x\in X}$ is a group structure if and only if $(H, X, G_x)_{x\in X}$ is a group structure. □

REMARK 2.7. *Inverse limit of group structures.* Let $\mathbf{G}_i = (G_i, X_i, G_{i,x})_{x\in X_i}$, $i \in I$, be an inverse system of group structures with connecting homomorphisms $\pi_{ji}\colon \mathbf{G}_j \to \mathbf{G}_i$. Put $G = \varprojlim G_i$, $X = \varprojlim X_i$, and let π_i be the projections on the ith coordinate of G and X. Since the π_{ji}'s commute with the action of G_i on X_i, they define a continuous action of G on X.

Next observe that $\mathrm{Subgr}(G) = \varprojlim \mathrm{Subgr}(G_i)$ as sets. Moreover, the étale topology of $\mathrm{Subgr}(G)$ coincides with the inverse limit of the étale topologies of $\mathrm{Subgr}(G_i)$. Indeed, let $H \in \mathrm{Subgr}(G)$. Each open neighborhood of H in the inverse limit of the étale topologies contains a set of the form $\mathrm{Subgr}(\pi^{-1}(U_i))$ for some $i \in I$ and $U_i \in \mathrm{Open}(G_i)$ with $H \le \pi^{-1}(U_i)$. This set is étale open in $\mathrm{Subgr}(G)$ because $\pi^{-1}(U_i)$ is open in G. Conversely, consider an open subgroup U of G containing H. For each $i \in I$ put $H_i = \pi_i(H)$ and $\mathcal{U}_i = \{U_i \in \mathrm{Open}(G_i) \mid H_i \le U_i\}$. Then $H_i = \bigcap_{U_i\in\mathcal{U}_i} U_i$ and $H = \bigcap_{i\in I} \pi^{-1}(H_i)$, hence $H = \bigcap_{i\in I}\bigcap_{U_i\in\mathcal{U}_i} \pi^{-1}(U_i)$. By compactness there are $i \in I$ and $U_i \in \mathcal{U}_i$ with $\pi^{-1}(U_i) \le U$. They satisfy, $\pi_i\big(\mathrm{Subgr}(\pi_i^{-1}(U_i))\big) \subseteq \mathrm{Subgr}(U_i)$. Hence, $\pi_i^{-1}\big(\mathrm{Subgr}(U_i)\big) \subseteq \mathrm{Subgr}\big(\pi_i^{-1}(U_i)\big) \subseteq \mathrm{Subgr}(U)$.

Since the π_{ji}'s commute with the maps $\delta_{\mathbf{G}_i}\colon X_i \to \mathrm{Subgr}(G_i)$, they define an map $\delta\colon X \to \mathrm{Subgr}(G)$ which is continuous in the inverse of the étale topologies of $\mathrm{Subgr}(G_i)$. By the preceding paragraph, δ is étale continuous. Specifically, for each $x = (x_i)_{i\in I}$ in X we have $\delta(x) = G_x = \varprojlim G_{i,x_i}$.

Since each $\delta_{\mathbf{G}_i}$ commutes with the action of $\mathbf{G}_i$, the map δ commutes with the action of G. Finally, with $x = (x_i)_{i\in I}$, it follows from $S_{x_i} \le G_{x_i}$ that $S_x \le G_x$. Therefore, $\mathbf{G} = (G, X, G_x)_{x\in X}$ is a group structure.

If each π_{ji} is rigid, then each π_i is rigid. If each π_{ji} is a cover, then so is each π_i. Indeed, Let $x = (x_k)_{k\in I}$ and $y = (y_k)_{k\in I}$ be elements of X satisfying $x_i = y_i$. Then, for each $j \ge i$ the closed subset $K_j = \{\kappa \in \mathrm{Ker}(\pi_{ji}) \mid x_j^\kappa = y_j\}$ of G_j is not empty. If $k \ge j$, then $\pi_{kj}(K_k) \subseteq K_j$. Therefore, there is a $\kappa \in G$ with $\pi_j(\kappa) \in K_j$ for all $j \ge i$. This κ belongs to $\mathrm{Ker}(\pi_i)$ and $x^\kappa = y$, as claimed. □

LEMMA 2.8. *Let $\mathbf{G} = (G, X, G_x)_{x\in X}$ be a group structure and $\mathcal{N}$ an inductive collection of closed normal subgroups of G with $\bigcap_{N\in\mathcal{N}} N = 1$. Then $\mathbf{G} = \varprojlim \mathbf{G}/N$ where N ranges over $\mathcal{N}$.*

PROOF. The only point which is perhaps not clear is $X = \varprojlim X/N$. To prove this equality define a map $f\colon X \to \varprojlim X/N$ by $f(x) = (x^N)_{N\in\mathcal{N}}$, where $x^N = \{x^\nu \mid \nu \in N\}$. Then f is continuous. Compactness of X implies f is surjective.

Since both X and $\varprojlim X/N$ are profinite spaces, it suffices now to prove that f is injective.

Consider distinct elements $x, y \in X$. Choose disjoint open subsets U and V of X with $x \in U$ and $y \in V$. Since the action of G on X is continuous, x has an open neighborhood U_0 and there is an $N \in \mathcal{N}$ with $U_0^N \subseteq U$. Then $x^\nu \notin V$, so $x^\nu \neq y$ for all $\nu \in N$. Therefore, $f(x) \neq f(y)$. □

CONSTRUCTION 2.9. *Fiber products.* Let

$$\mathbf{A} = (A, I, A_i)_{i\in I}, \quad \mathbf{B} = (B, J, B_j)_{j\in J}, \text{ and } \mathbf{G} = (G, X, G_x)_{x\in X}$$

be group structures. Let $\alpha: \mathbf{B} \to \mathbf{A}$ and $\varphi: \mathbf{G} \to \mathbf{A}$ be morphisms of group structures. Put

(5a) $H = B \times_A G = \{(b, g) \in B \times G \mid \alpha(b) = \varphi(g)\}$,
(5b) $Y = J \times_I X = \{(j, x) \in J \times X \mid \alpha(j) = \varphi(x)\}$, and
(5c) $H_y = B_j \times_A G_x = \{(b, g) \in B_j \times G_x \mid \alpha(b) = \varphi(g)\}$ for $y = (j, x) \in Y$.

Define a continuous action of H on Y by $(j, x)^{(b,g)} = (j^b, x^g)$. We claim: $\mathbf{H} = (H, Y, H_y)_{y\in Y}$ is a group structure.

To verify the claim it suffices to prove that the map $y \mapsto H_y$ is étale continuous. Indeed, let $y = (j, x)$. Consider an open subgroup W of H which contains $H_y = B_j \times_A G_x$. Let $\mathcal{U}$ be the set of all open subgroups of B which contain B_j. Let $\mathcal{V}$ be the set of all open subgroups of G which contain G_x. The intersection of all $U \times_A V$ with $U \in \mathcal{U}$ and $V \in \mathcal{V}$ is $B_j \times_A G_x$. Since $H \smallsetminus W$ is closed, there are an open subgroup U of B and an open subgroup V of G with $H_y \leq U \times_A V \leq W$. The set $Y_0 = \{(j', x') \in Y \mid B_{j'} \leq U,\ G_{x'} \leq V\}$ is an open neighborhood of y in Y and $H_{(j',x')} \leq W$ for each $(j', x') \in Y_0$. Therefore, the above map is continuous.

Finally let $\beta: H \to B$, $\beta: Y \to J$, $\psi: H \to G$, and $\psi: Y \to X$ be the projections on the coordinates. Then the following diagram of group structures is commutative:

(6)
$$\begin{array}{ccc} \mathbf{H} & \xrightarrow{\psi} & \mathbf{G} \\ {\scriptstyle\beta}\downarrow & & \downarrow{\scriptstyle\varphi} \\ \mathbf{B} & \xrightarrow{\alpha} & \mathbf{A} \end{array}$$

If both $\mathbf{B}$ and $\mathbf{G}$ are finite, then so is $\mathbf{H}$. □

DEFINITION 2.10. *Cartesian squares.* Let (6) be a commutative diagram of group structures. Call (6) a **cartesian square** if this holds: For all group structures $\mathbf{F}$ and morphisms $\beta': \mathbf{F} \to \mathbf{B}$ and $\psi': \mathbf{F} \to \mathbf{G}$ with $\alpha \circ \beta' = \varphi \circ \psi'$ there is a unique morphism $\epsilon: \mathbf{F} \to \mathbf{H}$ satisfying $\beta \circ \epsilon = \beta'$ and $\psi \circ \epsilon = \psi'$. □

LEMMA 2.11. *Let (6) be a commutative diagram of group structures.*

(a) *Suppose* $\mathbf{H} = \mathbf{B} \times_{\mathbf{A}} \mathbf{G}$ *and* β, ψ *are the coordinate projections. Then (6) is a cartesian square.*

(b) *Suppose (6) is a cartesian square. Put* $\mathbf{H}' = \mathbf{B} \times_{\mathbf{A}} \mathbf{G}$. *Let* $\psi': \mathbf{H}' \to \mathbf{B}$ *and* $\beta': \mathbf{H}' \to \mathbf{G}$ *be the projection maps. Then there is a unique isomorphism* $\gamma: \mathbf{H}' \to \mathbf{H}$ *with* $\psi \circ \gamma = \psi'$ *and* $\beta \circ \gamma = \beta'$.

PROOF. Statement (a) follows from the definition of $\mathbf{B} \times_{\mathbf{A}} \mathbf{G}$. Statement (b) follows from (a) and from the uniqueness of ϵ in Definition 2.10. □

LEMMA 2.12. *Suppose (6) is a cartesian square of group structures. Then:*
(a) $\beta\colon \mathrm{Ker}(\psi) \to \mathrm{Ker}(\alpha)$ *is an isomorphism.*
(b) *For each* $y \in Y$, $\psi\colon H_y \to G_{\psi(y)}$ *is injective if and only if* $\alpha\colon B_{\beta(y)} \to A_{\alpha(\beta(y))}$ *is injective.*
(c) *If* α *is a cover, then* ψ *is a cover.*
(d) *If* ψ *is a cover and* φ *is an epimorphism, then* α *is a cover.*

PROOF. By Lemma 2.11(b) we may assume that $\mathbf{H}$ is $\mathbf{B} \times_{\mathbf{A}} \mathbf{G}$ and β, ψ are the projections.

PROOF OF (a) AND (b). By assumption, $\mathrm{Ker}(\psi) = \mathrm{Ker}(\alpha) \times \{1\}$, which gives (a). Similarly, for $y = (j, x) \in Y$, (5c) and (a) imply that β maps $\mathrm{Ker}(\psi) \cap H_y$ isomorphically onto $\mathrm{Ker}(\alpha) \cap B_j$. This gives (b).

PROOF OF (c). Suppose α is a cover. Then $\alpha(B) = A$. Hence, for each $g \in G$ there is a $b \in B$ with $\alpha(b) = \varphi(g)$. Therefore, $(b, g) \in H$ and $\psi(b, g) = g$. Thus, $\psi(H) = G$. Similarly, $\psi(Y) = X$ and $\psi(H_y) = G_{\psi(y)}$ for each $y \in Y$. Since $\alpha\colon B_{\beta(y)} \to A_{\alpha(\beta(y))}$ is an isomorphism, (b) implies $\psi\colon H_y \to G_{\psi(y)}$ is an isomorphism.

Finally, suppose $\psi(j, x) = \psi(j', x')$. Then $x = x'$ and $\alpha(j) = \alpha(j')$. The rigidity of α gives $b \in \mathrm{Ker}(\alpha)$ with $j' = j^b$. Then $(b, 1) \in \mathrm{Ker}(\psi)$ and $(j', x') = (j, x)^{(b,1)}$. This proves ψ is a cover.

PROOF OF (d). By assumption, $\alpha(\beta(H)) = \varphi(\psi(H)) = A$ and $\alpha(\beta(Y)) = \varphi(\psi(Y)) = I$. Hence, $\alpha(B) = A$ and $\alpha(J) = I$.

Now let $j \in J$ and $i = \alpha(j)$. Since φ is an epimorphism, there is an $x \in X$ with $\varphi(x) = i$ and $\varphi(G_x) = A_i$. Put $y = (j, x)$. Since ψ is a cover, $\psi\colon H_y \to G_x$ is an isomorphism. Hence, $A_i \geq \alpha(B_j) \geq \alpha(\beta(H_y)) = \varphi(\psi(H_y)) = A_i$, so $\alpha(B_j) = A_i$. We conclude from (b) that $\alpha\colon B_j \to A_i$ is an isomorphism.

Finally, consider $j, j' \in J$ with $\alpha(j) = \alpha(j')$. Choose $x \in X$ with $\alpha(j) = \alpha(j') = \varphi(x)$. Then $\psi(j, x) = \psi(j', x)$. Hence, there is a $(b, 1) \in \mathrm{Ker}(\psi)$ with $(j', x) = (j, x)^{(b,1)}$. Therefore, $b \in \mathrm{Ker}(\alpha)$ and $j' = j^b$. This proves α is a cover. $\square$

3. Completion of a Cover to a Cartesian Square

There are several places in this work where a group structure $\mathbf{G} = (G, X, G_x)_{x \in X}$ is given and we need to define a morphism $f\colon (X, G) \to (Y, H)$ of group spaces, where $f\colon G \to H$ is a given homomorphism. If the set of G-orbits of X has a closed system of representatives X' (also called a **fundamental domain**), then $X \cong X' \times G$. Hence, we may first define f on X' and then extend it to X by the rule

$$f(x^\sigma) = f(x)^{f(\sigma)}, \qquad x \in X',\ \sigma \in G. \tag{1}$$

This could considerably simplify the proof of the Main Theorem. Unfortunately, fundamental domains do not always exist. (One may find a counter example of J. L. Kelly on page 473 of [ArK].) Instead we produce a "special partition" of $\mathbf{G}$ giving rise to a subset X' of X that "approximates" a fundamental domain in a way that allows the definition of the desired function f.

Lemmas 3.1, 3.2, 3.3, and 3.4 below prepare ingredients of the construction of special partitions in Lemma 3.6. The definition of "special partition" appears in Lemma 3.5. It follows by a specification of the above mentioned set X'.

LEMMA 3.1. *Let G be a profinite group acting continuously on a compact Hausdorff space X. Then:*

(a) *S_x is a closed subgroup of G.*

(b) *The map $x \mapsto S_x$ from X to* Subgr(G) *is étale continuous.*

PROOF OF (a). The action $a: X \times G \to X$ and the projection $p: X \times G \to X$ are continuous, so $\{x\} \times S_x = p^{-1}(x) \cap a^{-1}(x)$ is closed in $X \times G$. Therefore, S_x is closed in G.

PROOF OF (b). Let N be an open normal subgroup of G and let $x \in X$. We have to find an open neighborhood V of x with $S_y \le S_x N$ for all $y \in V$.

CASE A. *G is finite and $N = 1$.* Consider $\sigma \in G \setminus S_x$. Then, $x^\sigma \neq x$. Since X is Hausdorff, it has disjoint open subsets U_1, U_2 with $x \in U_1$ and $x^\sigma \in U_2$. Then $V_\sigma = U_1 \cap U_2^{\sigma^{-1}}$ is an open neighborhood of x. If $y \in V_\sigma$, then $y \in U_1$ and $y^\sigma \in U_2$, so $y^\sigma \neq y$. Since G is finite, $V = \bigcap_{\sigma \in G \setminus S_x} V_\sigma$ is open. Each $y \in V$ satisfies $S_y \le S_x$.

CASE B. *The general case.* The quotient space $\bar{X} = X/N$ is Hausdorff [Bre, Thm. 3.1(1)] and $\bar{G} = G/N$ acts continuously on $\bar{X}$. Use a bar for reduction modulo N. Case A gives an open neighborhood $\bar{V}$ of $\bar{x}$ in $\bar{X}$ with $S_{\bar{y}} \le S_{\bar{x}}$ for each $\bar{y} \in \bar{V}$. Then the preimage V of $\bar{V}$ in X is an open neighborhood of x in X. For each $y \in V$ we have $S_{\bar{y}} = S_y N/N$. Hence, $S_y \le S_x N$. □

LEMMA 3.2. *Let Y be a profinite space and A, B disjoint closed subsets. Then there are disjoint open-closed subsets U, V with $A \subseteq U$ and $B \subseteq V$.*

PROOF. As a profinite space, Y is compact and Hausdorff. Hence, it has disjoint open subsets U', V' with $A \subseteq U'$ and $B \subseteq V'$. The set U' is a union of open-closed subsets. Since A is compact, finitely many of them cover A. Their union U is an open-closed subset satisfying $A \subseteq U \subseteq U'$. Similarly, Y has an open-closed subset V with $B \subseteq V \subseteq V'$. It satisfies $U \cap V = \emptyset$. □

LEMMA 3.3. *Let (X, G) be a profinite group space, $x \in X$, and V an open neighborhood of x. Suppose $x^G \subseteq V$. Then x has an open-closed G-invariant neighborhood W with $W \subseteq V$.*

PROOF. Denote the images of points and subsets of X under the quotient map $\pi: X \to X/G$ by a bar. Since $F = X \setminus V$ is closed in X, $\bar{F}$ is closed in X/G. Moreover, $\bar{x} \notin \bar{F}$. Lemma 3.2 gives an open-closed subset $\bar{W}$ with $\bar{x} \in \bar{W}$ and $\bar{W} \cap \bar{F} = \emptyset$. Put $W = \pi^{-1}(\bar{W})$. Then W is open-closed in X, invariant under G, and $x \in W \subseteq V$. □

LEMMA 3.4. *Let (X, G) be a profinite group space, $x \in X$, and H an open subgroup of G. Suppose $S_x \le H$. Write $G = \dot{\bigcup}_{\rho \in R} H\rho$. Then x^H has an H-invariant open-closed neighborhood U satisfying $U^G = \dot{\bigcup}_{\rho \in R} U^\rho$.*

PROOF. The closed sets $x^{H\rho}$, $\rho \in R$, are disjoint, because $S_x \le H$. Hence, X has open disjoint sets V_ρ satisfying $x^{H\rho} \subseteq V_\rho$, $\rho \in R$. For each $\rho \in R$ we have $x^H \subseteq V_\rho^{\rho^{-1}}$. By Lemma 3.3, with H replacing G, there is an H invariant open-closed set U_ρ with $x^H \subseteq U_\rho \subseteq V_\rho^{\rho^{-1}}$.

Now consider the H-invariant open-closed set $U = \bigcap_{\rho \in R} U_\rho$. It satisfies $U^\rho \subseteq V_\rho$ for each $\rho \in R$, so the U^ρ are disjoint. Therefore, $U^G = \dot\bigcup_{\rho \in R} U^\rho$. □

DEFINITION 3.5. *Special partition.* Let $\mathbf{G} = (G, X, G_x)_{x \in X}$ be a group structure. A **special partition** of $\mathbf{G}$ is a data $(G_i, X_i, R_i)_{i \in I_0}$ satisfying these conditions:

(2a) I_0 is a finite set disjoint from X.
(2b) X_i is a nonempty open-closed subset of X, $i \in I_0$.
(2c) G_i is an open subgroup of G containing G_x for all $x \in X_i$ and $i \in I_0$.
(2d) $G_i = \{\sigma \in G \mid X_i^\sigma = X_i\}$, $i \in I_0$.
(2e) R_i is a finite subset of G and $G = \dot\bigcup_{\rho \in R_i} G_i \rho$, $i \in I_0$.
(2f) $X = \dot\bigcup_{i \in I_0} \dot\bigcup_{\rho \in R_i} X_i^\rho$.

Here is a consequence of (2a)-(2f):

(2g) Suppose $i, j \in I_0$ and $X_i^\sigma \cap X_j \neq \emptyset$. Then, $i = j$ and $\sigma \in G_i$.

To prove (2g) write $\sigma = \zeta\rho$ with $\zeta \in G_i$ and $\rho \in R_i$. By (2d), $X_i^\rho \cap X_j \neq \emptyset$. Hence, by (2f), $i = j$ and $X_i^\rho = X_i$. By (2d), $\rho \in G_i$. Therefore, $\sigma \in G_i$.

By (2d), each R_i can be replaced by every set R_i' satisfying $G = \dot\bigcup_{\rho \in R_i'} G_i \rho$. Thus, we also call $(G_i, X_i)_{i \in I_0}$ a **special partition** of $\mathbf{G}$ if there exist R_i, $i \in I_0$ satisfying (2a)-(2f). In this case every system $(R_i)_{i \in I_0}$ satisfying (2e) also satisfies (2f). □

Now suppose $\mathbf{G} = (G, X, G_x)_{x \in X}$ is a group structure, (Y, H) is a profinite group space, and $\varphi: G \to H$ is a homomorphism which we wish to extend to a morphism $\varphi: (X, G) \to (Y, H)$ of profinite group spaces. We construct a special partition $(X_i, G_i, R_i)_{i \in I_0}$ of $\mathbf{G}$ such that φ has a natural definition on $X' = \bigcup_{i \in I_0} X_i$ satisfying $\varphi(x) = \varphi(x)^{\varphi(\sigma)}$ for all $x \in X'$ and $\sigma \in G$ with $x^\sigma \in X'$. Then $\varphi(x^\tau) = \varphi(x)^{\varphi(\tau)}$ for arbitrary $\tau \in G$ will define the desired extension φ.

This procedure allows us to extend each epimorphism φ of G onto a finite group A to an epimorphisms of $\mathbf{G}$ onto a finite group structure $\mathbf{A} = (A, I, A_i)_{i \in I}$ (Lemma 3.7). Consequently, each cover $\psi: \mathbf{H} \to \mathbf{G}$ of group structures with a finite kernel can be completed to a cartesian square as in (6) of Section 2 such that $\alpha: \mathbf{B} \to \mathbf{A}$ is a cover of finite group structures (Lemma 3.9). The latter result is the main ingredient in the transition from solving finite embedding problems to solving arbitrary embedding problems of projective group structures (Proposition 4.2).

LEMMA 3.6. *Let $\mathbf{G} = (G, X, G_x)_{x \in X}$ be a group structure, Y be a subset of X, and Y_0 a finite subset of Y. Suppose $X = Y^G$ and the elements of Y_0 belong to distinct G-orbits. For each $y \in Y$ let G_y' be an open subgroup of G containing G_y and let V_y be an open neighborhood of $y^{G_y'}$ in X. Then there exists a finite subset $\{y_i \mid i \in I_0\}$ of Y containing Y_0 and a special partition $(G_{y_i}', X_i)_{i \in I_0}$ of $\mathbf{G}$ such that $y_i \in X_i \subseteq V_{y_i}$ for all $i \in I_0$.*

PROOF. We may assume Y is a (not necessarily closed) system of representatives of the G-orbits of X. For each $y \in Y$ use Lemma 3.4 to replace V_y by another set, if necessary, to assume:

(3a) V_y is open-closed, G_y'-invariant, and $y^{G_y'} \subseteq V_y$.
(3b) Writing $G = \dot\bigcup_{\rho \in R_y} G_y' \rho$, we have $V_y^G = \dot\bigcup_{\rho \in R_y} V_y^\rho$.

The rest of the proof has three parts.

PART A. *Finite covering of X.* By assumption, $X = \bigcup_{y\in Y} y^G \subseteq \bigcup_{y\in Y} V_y^G$. Hence, by compactness, there is a finite subset $\{y_i \mid i \in I_0\}$ of Y with $X = \bigcup_{i\in I_0} V_{y_i}^G$. Add the elements of Y_0 to $\{y_i \mid i \in I_0\}$, if necessary, to assume that $Y_0 \subseteq \{y_i \mid i \in I_0\}$. By our choice of Y, the sets y_i^G, $i \in I_0$, are closed and disjoint. Hence, there are disjoint open subsets W_i' with $y_i^G \subseteq W_i'$, $i \in I_0$. For each $i \in I_0$ Lemma 3.3 gives a G-invariant open-closed set W_i with $y_i^G \subseteq W_i \subseteq V_{y_i}^G \cap W_i'$.

PART B. *Making V_{y_i} smaller.* By Part A,

$$y_i \in W_i \smallsetminus \bigcup_{j\neq i} W_j \subseteq V_{y_i}^G \smallsetminus \bigcup_{j\neq i} W_j$$

and $\bigcup_{j\neq i} W_j$ is G-invariant. Let $V_i = V_{y_i} \smallsetminus \bigcup_{j\neq i} W_j$. Then V_i is a G'_{y_i}-invariant open-closed set which, by (3), satisfies

(4) $V_i^G = \dot\bigcup_{\rho\in R_i} V_i^\rho$

where $R_i = R_{y_i}$. Moreover, $y_i \in V_i \smallsetminus \bigcup_{j\neq i} V_j^G$. Indeed $y_i \in W_i$. If $y_i \in V_j^G$ for $j \neq i$, then there is a $\sigma \in G$ with $y_i^\sigma \in V_j$, so $y_i^\sigma \notin W_i$. But W_i is G-invariant. Hence, $y_i \notin W_i$, a contradiction.

We claim that $X = \bigcup_{i\in I_0} V_i^G$. Indeed, let $x \in X$. If there is an i with $x \in W_i$, then $x \notin \bigcup_{j\neq i} W_j$. Hence, $x \in V_i^G$. Else, $x \notin \bigcup_{j\in I_0} W_j$ and there is an i with $x \in V_{y_i}^G$ (Part A). Therefore, $x \in V_{y_i}^G \smallsetminus \bigcup_{j\neq i} W_j = V_i^G$.

PART C. *Separating V_i.* Let $X_i = V_i \smallsetminus \bigcup_{j=1}^{i-1} V_j^G$, $i \in I_0$. Then $X = \dot\bigcup_{i\in I_0} X_i^G$. Also, X_i is a G'_{y_i}-invariant open-closed neighborhood of y_i and $X_i \subseteq V_{y_i}$. By (4), $X_i^G = \dot\bigcup_{\rho\in R_i} X_i^\rho$.

Finally, consider $\sigma \in G$ with $X_i^\sigma = X_i$. Write $\sigma = \zeta\rho$ with $\zeta \in G'_{y_i}$ and $\rho \in R_i$. Then $X_i = X_i^\sigma = X_i^\rho$. By the preceding paragraph, $\rho \in G'_{y_i}$. Therefore, $\sigma \in G'_{y_i}$.□

LEMMA 3.7. *Let $\mathbf{G} = (G, X, G_x)_{x\in X}$ be a group structure, A a finite group, and $\varphi: G \to A$ an epimorphism. Then:*

(a) *φ extends to an epimorphism φ of $\mathbf{G}$ onto a finite group structure $\mathbf{A} = (A, I, A_i)_{i\in I}$.*
(b) *Let X_0 be a finite subset of X. Then φ may be constructed in (a) with $\varphi(G_x) = A_{\varphi(x)}$ for each $x \in X_0$.*
(c) *Suppose $X = \dot\bigcup_{j\in J} Y_j$ with J finite, each Y_j is open-closed, G permutes the Y_j's, and $Y_j^\nu = Y_j$ for all $j \in J$ and $\nu \in \mathrm{Ker}(\varphi)$. Then φ may be constructed in (a) such that $\varphi(Y_j)$, $j \in J$, are disjoint.*
(d) *Let $y_1, \dots, y_m$ be elements of X lying in distinct G-orbits. Then φ may be constructed in (a) such that $\varphi(x_1), \dots, \varphi(x_n)$ lie in distinct A-orbits.*

PROOF OF (a). We may assume $A = G/N$ with $N = \mathrm{Ker}(\varphi)$. Both maps $x \mapsto G_x$ and $x \mapsto S_x$ of X into $\mathrm{Subgr}(G)$ are étale continuous (by definition and by Lemma 3.1). Hence, for each $y \in X$ the set $V_y = \{x \in X \mid S_x \le S_yN,\ G_x \le G_yN\}$ is open and contains $y^N = y^{S_yN}$. Lemma 3.6, with S_yN replacing G'_y, gives a finite subset $\{y_i \mid i \in I_0\}$ of X and a special partition $(S_{y_i}N, X_i)_{i\in I_0}$ of $\mathbf{G}$ such that

(5) $y_i \in X_i \subseteq V_{y_i}$ for all $i \in I_0$.

Thus, the following holds:

(6a) X_i is open closed in X, $i \in I_0$.
(6b) $S_{y_i}N = \{\sigma \in G \mid X_i^\sigma = X_i\}$.
(6c) $X = \dot\bigcup_{i\in I_0} \dot\bigcup_{\rho\in R_i} X_i^\rho$, where $G = \dot\bigcup_{\rho\in R_i} S_{y_i}N\rho$.

Set $I=\bigcup_{i\in I_0}\{X_i^\sigma \mid \sigma\in G\}=\bigcup_{i\in I_0}\{X_i^\rho \mid \rho\in R_i\}$. Since R_i are finite, I is finite and G acts on I from the right. For $i\in I_0$, $\sigma\in G$, and $\nu\in N$, (6b) implies $X_i^{\sigma\nu}=(X_i^{\sigma\nu\sigma^{-1}})^\sigma=X_i^\sigma$. Hence, the action of G induces an action of G/N on I.

Next define a map $\varphi\colon X\to I$ such that $\varphi(x)=X_i^\rho$ for all $i\in I_0$ and $\rho\in R_i$ and each $x\in X_i^\rho$. Since (6c) is a partition of X into open-closed sets, φ is surjective and continuous. Let $i\in I_0$, $y\in X_i$, and $\sigma\in G$. Write $\sigma=\tau\rho$ with $\tau\in S_{y_i}N$ and $\rho\in R_i$. Then $y^\sigma\in X_i^\rho$ (by (6b)) and $\varphi(y^\sigma)=X_i^\rho=X_i^\sigma$. Thus,

$$\varphi(y^\sigma)=X_i^\sigma \qquad \text{for } y\in X_i,\ \sigma\in G. \tag{7}$$

It follows that $\varphi(x^\sigma)=\varphi(x)^{\varphi(\sigma)}$ for all $x\in X$ and $\sigma\in G$.

For each $X_i^\sigma\in I$ put $A_{X_i^\sigma}=\varphi(G_{y_i}^\sigma)=\varphi(G_{y_i})^{\varphi(\sigma)}$. This is a good definition: If $X_i^\sigma=X_j^{\sigma'}$, then, by (6), $i=j$ and $\sigma'=\zeta\nu\sigma$ with $\zeta\in S_{y_i}\le G_{y_i}$ and $\nu\in N$. Then $\varphi(G_{y_j}^{\sigma'})=\varphi(G_{y_i}^\zeta)^{\varphi(\nu)\varphi(\sigma)}=\varphi(G_{y_i})^{\varphi(\sigma)}=\varphi(G_{y_i}^\sigma)$.

We claim that $\varphi(G_x)\le A_{\varphi(x)}$ for all $x\in X$. Indeed, there are $y\in X_i$ and $\sigma\in G$ such that $x=y^\sigma$. By (5), $y\in V_{y_i}$. Hence, $G_y\le G_{y_i}N$, so $\varphi(G_y)\le\varphi(G_{y_i})$. Therefore,

$$\varphi(G_x)=\varphi(G_y^\sigma)=\varphi(G_y)^{\varphi(\sigma)}\le\varphi(G_{y_i})^{\varphi(\sigma)}=A_{X_i^\sigma}=A_{\varphi(x)}.$$

Finally, by (6b), the stabilizer of $X_i\in I$ in G/N is contained in $S_{y_i}N/N=\varphi(S_{y_i})$. Therefore it is contained in $\varphi(G_{y_i})=A_{X_i}$. Consequently, $(G/N,I,A_i)_{i\in I}$ is a finite group structure.

PROOF OF (b). Let Y_0 be a subset of X_0 with $X_0\subseteq Y_0^G$ and $y^\sigma\ne y'$ for all distinct $y,y'\in Y_0$ and $\sigma\in G$. By Lemma 3.6 we may assume $\{y_i\mid i\in I_0\}$ contains Y_0. Write each $x\in X_0$ as $x=y^\sigma$ with $y\in Y_0$ and $\sigma\in G$. Then $y=y_i$ for some $i\in I_0$ and $\varphi(G_x)=\varphi(G_{y_i}^\sigma)=A_{X_i^\sigma}=A_{\varphi(x)}$.

PROOF OF (c). For each $y\in Y$ we may choose V_y at the beginning of the proof of (a) such that V_y is contained in the unique Y_j which contains y^N. By (5), each X_i with $i\in I_0$ is contained in a unique Y_j with $j\in J$. Since G permutes the Y_j's, each X_i^ρ with $\rho\in R_i$ is contained in a unique Y_j with $j\in J$. Hence, $Y_j=\bigcup_{(i,\rho)\in S_j}X_i^\rho$ with disjoint subsets S_j of $\{(i,\rho)\mid i\in I_0,\ \rho\in R_i\}$. Therefore, $\varphi(Y_j)=\{A_{X_i^\rho}\mid(i,\rho)\in S_j\}$ are disjoint.

PROOF OF (d). Lemma 3.6 allows us to choose I_0 at the beginning of the proof of (a) such that $\{1,\dots,m\}\subseteq I_0$. By (7), $\varphi(x_i)=A_{X_i}$ belong then to distinct A-orbits. □

Lemma 3.7 has several consequences.

LEMMA 3.8. *Let $\varphi\colon\mathbf{G}\to\mathbf{A}$ be a morphism of group structures with $\mathbf{A}$ finite and N_0 an open subgroup of the underlying group G. Then there are a morphism $\bar\varphi\colon\hat{\mathbf{A}}\to\mathbf{A}$ of group structures and an epimorphism $\hat\varphi\colon\mathbf{G}\to\hat{\mathbf{A}}$ satisfying $\varphi=\bar\varphi\circ\hat\varphi$ and* $\mathrm{Ker}(\hat\varphi)\le N_0$.

PROOF. Let $\mathbf{G}=(G,X,G_x)_{x\in X}$ and $\mathbf{A}=(A,I,A_i)_{i\in I}$. For each $i\in I$ let $X_i=\varphi^{-1}(i)$. Then G permutes the finite set $\{X_i\mid i\in I\}$. Hence, G has an open normal subgroup N such that $N\le N_0\cap\mathrm{Ker}(\varphi)$ and $X^\nu=X$ for each $\nu\in N$ and $i\in I$.

Set $\hat{A} = G/N$ and let $\hat{\varphi}: G \to \hat{A}$ be the quotient map. Use Lemma 3.7 to extend $\hat{\varphi}: G \to \hat{A}$ to an epimorphism $\hat{\varphi}: \mathbf{G} \to \hat{\mathbf{A}}$ such that $\hat{\mathbf{A}} = (\hat{A}, J, \hat{A}_j)_{j\in J}$ is finite and $\hat{\varphi}(X_i)$, $i \in I$, are disjoint.

Now define $\bar{\varphi}: \hat{A} \to A$ to be the map induced by φ. Define $\bar{\varphi}: J \to I$ by $\bar{\varphi}(j) = i$ for all $j \in \hat{\varphi}(X_i)$ and $i \in I$. Then $\bar{\varphi}: \hat{\mathbf{A}} \to \mathbf{A}$ is a morphism of finite group structures and $\varphi = \bar{\varphi} \circ \hat{\varphi}$. □

LEMMA 3.9. *Let $\psi: \mathbf{H} \to \mathbf{G}$ be a cover of group structures with a finite kernel. Then there is a cartesian square of group structures*

$$\begin{array}{ccc} \mathbf{H} & \xrightarrow{\psi} & \mathbf{G} \\ \Big\downarrow{\scriptstyle\beta} & & \Big\downarrow{\scriptstyle\varphi} \\ \mathbf{B} & \xrightarrow{\alpha} & \mathbf{A} \end{array} \tag{8}$$

in which $\mathbf{A}$ and $\mathbf{B}$ are finite and α is a cover.

PROOF. Let $\mathbf{G} = (G, X, G_x)_{x\in X}$ and $\mathbf{H} = (H, X, H_y)_{y\in Y}$. By Lemma 2.3, $\bigcup_{y\in Y} H_y$ is a closed subset of H. By assumption, $K = \mathrm{Ker}(\psi)$ is a finite group and $\bigcup_{y\in Y} H_y \cap (K \smallsetminus 1) = \emptyset$. Hence, H has an open normal subgroup N with $\left(\bigcup_{y\in Y} H_y\right)N \cap (K \smallsetminus 1) = \emptyset$. Thus, $N \cap K = 1$ and $H_yN \cap KN = N$ for each $y \in Y$.

Let $B = H/N$ and let $\beta: H \to B$ be the quotient map. Use Lemma 3.7 to complete β to an epimorphism $\beta: \mathbf{H} \to \mathbf{B}$ with $\mathbf{B} = (B, J, B_j)_{j\in J}$ a finite group structure.

By Example 2.5, we may assume ψ is the quotient map $\mathbf{H} \to \mathbf{H}/K$. Put $\mathbf{A} = (A, I, A_i)_{i\in I} = \mathbf{B}/\beta(K)$. Then let $\alpha: \mathbf{B} \to \mathbf{A}$ be the quotient map and $\varphi: \mathbf{G} \to \mathbf{A}$ the epimorphism which β induces. This gives the commutative diagram (8). The assumption $H_yN \cap KN = N$ implies $B_j \cap \mathrm{Ker}(\alpha) = 1$ for each $j \in J$. Hence, by Example 2.5, α is a cover.

To prove (8) is cartesian, it suffices to check that the unique morphism $\epsilon: \mathbf{H} \to \mathbf{B} \times_{\mathbf{A}} \mathbf{G}$ induced by β and ψ is an isomorphism. Indeed, the group homomorphism $\epsilon: H \to B \times_A G$ is an isomorphism [FrJ, Section 22.2]. We show $\epsilon: Y \to J \times_I X$ is a bijection (hence, a homeomorphism): Let $(j, x) \in J \times_I X$. There is a $y \in Y$ such that $\psi(y) = x$. Since $\alpha(\beta(y)) = \varphi(x) = \alpha(j)$, there is a unique $b \in \mathrm{Ker}(\alpha) = \beta(K)$ with $\beta(y)^b = j$. Choose $k \in K$ with $\beta(k) = b$. Then $\beta(y^k) = j$ and $\psi(y^k) = \psi(y) = x$. Hence, $\epsilon(y^k) = (j, x)$. Therefore, ϵ is surjective.

Next let $y, y' \in Y$ with $\epsilon(y) = \epsilon(y')$. Then $\beta(y) = \beta(y')$ and $\psi(y) = \psi(y')$. Since ψ is a cover, there is a $k \in K$ with $y^k = y'$. Hence, $\beta(y)^{\beta(k)} = \beta(y') = \beta(y)$. Hence, $\beta(k) \in \beta(K) \cap S_{\beta(y)} \le \beta(K) \cap B_{\beta(y)} = 1$ (because $\mathrm{Ker}(\alpha) = \beta(K)$ and α is a cover). Thus, $k \in N \cap K = 1$. Therefore, $y = y'$. We conclude that ϵ is injective, hence bijective.

Since α is a cover, Lemma 2.12(c) implies that the projection $\psi': \mathbf{B} \times_{\mathbf{A}} \mathbf{G} \to \mathbf{G}$ is a cover. By assumption ψ is a cover. Hence, for each $y \in Y$ and $(j, x) = \epsilon(y)$ both $\psi': B_j \times_A G_x \to G_x$ and $\psi: H_y \to G_x$ are isomorphisms. Also, $\epsilon(H_y) \le B_j \times_A G_x$. Since $\psi = \psi \circ \epsilon$, the map $\epsilon: H_y \to B_j \times_A G_x$ is an isomorphism. This concludes the proof that (8) is cartesian. □

4. Projective Group Structures

The notion "projective group structure" which we introduce here replaces the notion "relatively projective group" of [HaJ3, Def. 4.2], also called "strongly relatively projective" in [Pop, p. 4]. The projective group structure is one of the two main objects which we put in duality in this work, the other one being "field-valuation structure with the block approximation condition" (Section 12).

Let $\mathbf{G}$ be a group structure. An **embedding problem** for $\mathbf{G}$ is a pair

$$(\varphi\colon \mathbf{G} \to \mathbf{A},\ \alpha\colon \mathbf{B} \to \mathbf{A}) \tag{1}$$

of morphisms of group structures in which α is a cover. A **solution** of (1) is a morphism $\gamma\colon \mathbf{G} \to \mathbf{B}$ with $\alpha \circ \gamma = \varphi$. The embedding problem is **finite** if $\mathbf{B}$ is finite. We say $\mathbf{G}$ is **projective**, if every finite embedding problem for $\mathbf{G}$ has a solution.

LEMMA 4.1. *Let $\mathbf{G}$ be a group structure. Suppose every finite embedding problem (1) for $\mathbf{G}$ where φ is an epimorphism is solvable. Then $\mathbf{G}$ is projective.*

PROOF. Lemma 3.8 gives a morphism $\bar{\varphi}\colon \hat{\mathbf{A}} \to \mathbf{A}$ of finite group structures and an epimorphism $\hat{\varphi}\colon \mathbf{G} \to \hat{\mathbf{A}}$ satisfying $\varphi = \bar{\varphi} \circ \hat{\varphi}$. Set $\hat{\mathbf{B}} = \mathbf{B} \times_{\mathbf{A}} \hat{\mathbf{A}}$. Let $\beta\colon \hat{\mathbf{B}} \to \mathbf{B}$ and $\hat{\alpha}\colon \hat{\mathbf{B}} \to \hat{\mathbf{A}}$ be the projection maps. By Lemma 2.12, $\hat{\alpha}\colon \hat{B} \to \hat{A}$ is a cover. Hence, $(\hat{\varphi}\colon \mathbf{G} \to \hat{\mathbf{A}},\ \hat{\alpha}\colon \hat{\mathbf{B}} \to \hat{\mathbf{A}})$ is a finite embedding problem. By assumption, there is a morphism $\hat{\gamma}\colon \mathbf{G} \to \hat{\mathbf{B}}$ with $\hat{\alpha} \circ \hat{\gamma} = \hat{\varphi}$. Then $\gamma = \beta \circ \hat{\gamma}$ is a solution of (1). Consequently, $\mathbf{G}$ is projective. □

Gruenberg proved that if every finite embedding problem for a profinite group G is weakly solvable, then every embedding problem for G is weakly solvable [FrJ, Lemma 22.3.2]. Gruenberg's proof goes through in the category of group structures almost verbatim.

PROPOSITION 4.2. *Let $\mathbf{G}$ be a projective group structure. Then every embedding problem for $\mathbf{G}$ has a solution.*

PROOF. Let (1) be an embedding problem for $\mathbf{G}$. Put $K = \mathrm{Ker}(\alpha)$.

PART A. *Suppose K is finite.* Lemma 3.9 gives a cartesian square of group structures

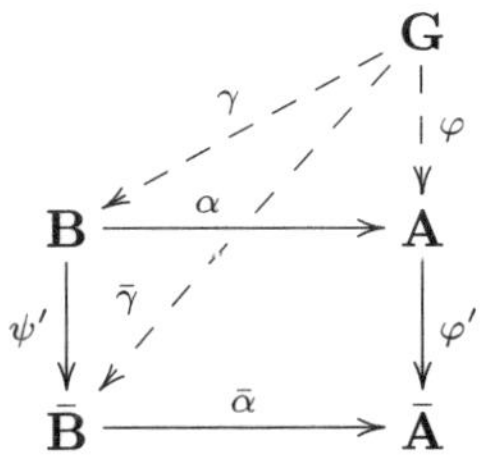

(without the dashed morphisms) in which $\bar{\mathbf{B}}$ and $\bar{\mathbf{A}}$ are finite, $\bar{\alpha}$ is a cover, and $\mathbf{B} = \bar{\mathbf{B}} \times_{\bar{\mathbf{A}}} \mathbf{A}$. Put $\bar{\varphi} = \varphi' \circ \varphi$. Then $(\bar{\varphi}\colon \mathbf{G} \to \bar{\mathbf{A}},\ \bar{\alpha}\colon \bar{\mathbf{B}} \to \bar{\mathbf{A}})$ is a finite embedding problem for $\mathbf{G}$. By assumption there is a morphism $\bar{\gamma}\colon \mathbf{G} \to \bar{\mathbf{B}}$ with $\bar{\alpha} \circ \bar{\gamma} = \bar{\varphi}$. Hence, there is a morphism $\gamma\colon \mathbf{G} \to \mathbf{B}$ with $\alpha \circ \gamma = \varphi$ and $\psi' \circ \gamma = \bar{\gamma}$ (Definition 2.10). In particular γ solves embedding problem (1).

PART B. *Application of Zorn's lemma.* Suppose (1) is an arbitrary embedding problem for $\mathbf{G}$. By Example 2.5 we may assume $\mathbf{A} = \mathbf{B}/K$ and α is the quotient map. For each closed normal subgroup L of B contained in K let $\alpha_L\colon \mathbf{B}/L \to \mathbf{A}$ be the quotient map $\mathbf{B}/L \to (\mathbf{B}/L)/(K/L)$. Then, α_L is a cover (Example 2.5) and

$$(\varphi\colon \mathbf{G} \to \mathbf{A},\ \alpha_L\colon \mathbf{B}/L \to \mathbf{A}) \tag{2}$$

is an embedding problem for $\mathbf{G}$. Let Λ be the set of pairs (L,γ) where L is a closed normal subgroup of B contained in K and γ is a solution of (2). The pair (K,φ) belongs to Λ. Partially order Λ by $(L',\gamma') \le (L,\gamma)$ if $L' \le L$ and $\alpha_{L',L} \circ \gamma' = \gamma$. Here $\alpha_{L',L}\colon \mathbf{B}/L' \to \mathbf{B}/L$ is the cover $\mathbf{B}/L' \to (\mathbf{B}/L')/(L/L')$.

Suppose $\Lambda_0 = \{(L_j,\gamma_j) \mid j \in J\}$ is a descending chain in Λ. Then $\varprojlim \mathbf{B}/L_j = \mathbf{B}/L$ with $L = \bigcap_{j\in J} L_j$ (Lemma 2.8). The γ_j's define a morphism $\gamma\colon \mathbf{G} \to \mathbf{B}/L$ with $\alpha_{L,L_j} \circ \gamma = \gamma_j$ for each $j \in J$. Thus, (L,γ) is a lower bound to Λ_0.

Zorn's lemma gives a minimal element (L,γ) of Λ. It suffices to prove that $L = 1$.

Assume $L \ne 1$. Then B has an open normal subgroup N with $L \not\le N$. Thus, $L' = N \cap L$ is a proper open subgroup of L which is normal in B. Then $(\gamma\colon \mathbf{G} \to \mathbf{B}/L,\ \alpha_{L',L}\colon \mathbf{B}/L' \to \mathbf{B}/L)$ is an embedding problem for $\mathbf{G}$. Its kernel $\mathrm{Ker}(\alpha_{L',L}) = L/L'$ is a finite group. Hence, by Part A, it has a solution γ'. The pair, (L',γ') is an element of Λ which is strictly smaller than (L,γ). This contradiction to the minimality of (L,γ) proves that $L = 1$, as desired. □

Corollary 4.3. *Let $\psi\colon \mathbf{H} \to \mathbf{G}$ be a cover of group structures. Suppose $\mathbf{G}$ is projective. Then $\mathbf{H}$ has a sub-group-structure $\mathbf{H}'$ which ψ maps isomorphically onto $\mathbf{G}$.*

Proof. Suppose $\mathbf{G} = (G,X,G_x)_{x\in X}$ and $\mathbf{H} = (H,Y,H_y)_{y\in Y}$. Proposition 4.2 gives a morphism $\gamma\colon \mathbf{G} \to \mathbf{H}$ with $\psi \circ \gamma = \mathrm{id}_{\mathbf{G}}$. Let $H' = \gamma(G)$ and $Y' = \gamma(X)$. Then $\psi\colon H' \to G$ is an isomorphism and $\psi\colon Y' \to X$ is a homeomorphism. Next, let $x \in X$ and $y' = \gamma(x)$. Then $\psi(y') = x$ and $\gamma(G_x) \le H_{y'}$. As a cover, ψ maps both $H_{y'}$ and $\gamma(G_x)$ isomorphically onto G_x. Hence, $\gamma(G_x) = H_{y'}$. In particular, $H_{y'} \le H'$. It follows that $y' \mapsto H_{y'}$ is a continuous map of Y' into $\mathrm{Subgr}(H')$ (Remark 2.6). Thus, $\mathbf{H}' = (H',Y',H_{y'})_{y'\in Y}$ is a sub-group-structure of $\mathbf{H}$ which ψ maps isomorphically onto $\mathbf{G}$. □

We shall have several occasions to use the following result of Herfort and Ribes.

Proposition 4.4 (Herfort-Ribes). *Let $G = \mathop{\ast}\limits_{i\in I} G_i$ be the free profinite product of finitely many profinite groups G_i. Then $G_i^g \cap G_j \ne 1$ implies $i = j$ and $g \in G_i$.*

Proof. The case $i = j$ is a combination of Proposition 2 and Theorem B' of [HeR]. The case $i \ne j$ cannot occur, otherwise the canonical map $\mathop{\ast}\limits_{k\in I} G_i \to \prod_{k\in I} G_i$ maps $G_i^g \cap G_i$ injectively onto 1. □

Lemma 4.5. *Let $\mathbf{A} = (A,I,A_i)_{i\in I}$ be a group structure, $\alpha\colon B \to A$ an epimorphism of profinite groups, and I_0 be a finite system of representatives of the A-orbits of I. For each $i \in I_0$ let B_i be a closed subgroup of B which α maps isomorphically onto A_i. Then α extends to a cover $\alpha\colon \mathbf{B} \to \mathbf{A}$, where $\mathbf{B} = (B,J,B_j)_{j\in J}$ is a group structure. Moreover, there is a map $\alpha'\colon I_0 \to J$ such that $J = \alpha'(I_0)^B$, $\alpha(\alpha'(i)) = i$, and $B_i = B_{\alpha'(i)}$ for each $i \in I_0$.*

PROOF. Consider $i \in I_0$. Then $S_i = \{a \in A \mid i^a = i\}$ is a closed subgroup of A_i. Hence, $T_i = \alpha^{-1}(S_i) \cap B_i$ is a closed subgroup of B_i which α maps bijectively onto S_i. Also, the set $\{(i, T_i b) \mid b \in B\}$ bijectively corresponds to the profinite quotient space B/T_i. Hence, $J = \bigcup_{i \in I_0} \{(i, T_i b) \mid b \in B\}$ is a profinite space. The rule $(i, T_i b)^{b'} = (i, T_i b b')$ defines a continuous action of B on J. For each $j = (i, T_i b) \in J$ let $B_j = B_i^b$. Then $j \mapsto B_j$ is a strictly continuous, (hence also étale continuous) map from J into $\mathrm{Subgr}(B)$.

Now suppose $(i, T_i b)^{b'} = (i, T_i b)$. Then $T_i b b' = T_i b$. Hence, $b' \in T_i^b \le B_i^b$. Therefore, $\mathbf{B} = (B, J, B_j)_{j \in J}$ is a group structure.

Next define a map $\alpha: J \to I$ by $\alpha(i, T_i b) = i^{\alpha(b)}$. If $\alpha(i', T_{i'} b') = \alpha(i, T_i b)$, then $i^{\alpha(b)} = (i')^{\alpha(b')}$. Since $i, i' \in I_0$, this implies $i = i'$ and $\alpha(b) = s_i \alpha(b')$ for some $s_i \in S_i$. Let t_i be the element of T_i with $\alpha(t_i) = s_i$. Then there is a $k \in \mathrm{Ker}(\alpha)$ with $b = t_i b' k$. Hence, $(i, T_i b) = (i, T_i t_i b' k) = (i', T_{i'} b')^k$. It follows, $\alpha: \mathbf{B} \to \mathbf{A}$ is a cover.

Finally define a map $\alpha': I_0 \to J$ by $\alpha'(i) = (i, T_i)$. Then $(i, T_i b) = \alpha'(i)^b$ for each $i \in I_0$ and $b \in B$, so $J = \alpha'(I_0)^B$. Also, $\alpha(\alpha'(i)) = \alpha(i, T_i) = i$ and $B_{\alpha'(i)} = B_{(i, T_i)} = B_i$ for each $i \in I_0$. $\square$

The assumption on a group structure $\mathbf{G}$ to be projective poses some restrictions on $\mathbf{G}$:

PROPOSITION 4.6. *Let $\mathbf{G} = (G, X, G_x)_{x \in X}$ be a projective group structure.*

(a) *Let $x, y \in X$ with $G_x \cap G_y \ne 1$. Then $y = x^g$ for some $g \in G_x$. Hence, $G_x = G_y$.*

(b) *Let $x \in X$ with $G_x \ne 1$. Then G_x is its own normalizer in G.*

(c) *Suppose $1 \notin \mathrm{StrictClosure}\{G_x \mid x \in X\}$ and $G_x = S_x$ for each $x \in X$. Then $\mathbf{G}$ is a proper structure.*

PROOF OF (a). There is an epimorphism $\bar{\varphi}: G \to \bar{A}$ with $\bar{A}$ finite and $\bar{\varphi}(G_x \cap G_y) \ne 1$. Consider an arbitrary epimorphism $\varphi: G \to A$ with A finite and $\mathrm{Ker}(\varphi) \le \mathrm{Ker}(\bar{\varphi})$. Then $\varphi(G_x \cap G_y) \ne 1$.

Use Lemma 3.7 to complete φ to an epimorphism $\varphi: \mathbf{G} \to \mathbf{A}$ of group structures with $\mathbf{A} = (A, I, A_i)_{i \in I}$ finite such that $\varphi(G_x) = A_{\varphi(x)}$ and $\varphi(x), \varphi(y)$ are not in the same A-orbit if x, y are not in the same G-orbit.

Assume without loss that I does not contain the symbol 0. Choose a system of representatives I_0 for the A-orbits of I which does not contain the symbol 0. Put $I_0' = \{0\} \cup I_0$ and $A_0 = A$. For each $i \in I_0'$ choose an isomorphic copy B_i of A_i and an isomorphism $\alpha_i: B_i \to A_i$.

Now consider the free profinite product $B = \prod^{*}_{i \in I_0'} B_i$. Let $\alpha: B \to A$ be the unique epimorphism with $\alpha|_{B_i} = \alpha_i$, $i \in I_0'$. Lemma 4.5 extends B to a group structure $\mathbf{B} = (B, J, B_j)_{j \in J}$ and α to a cover $\alpha: \mathbf{B} \to \mathbf{A}$. Moreover, there is a map $\alpha': I_0 \to J$ such that $J = \alpha'(I_0)^B$, $\alpha(\alpha'(i)) = i$, and $B_i = B_{\alpha'(i)}$ for each $i \in I_0$.

Since $\mathbf{G}$ is projective, Proposition 4.2 gives a morphism $\gamma: \mathbf{G} \to \mathbf{B}$ with $\alpha \circ \gamma = \varphi$. In particular $\alpha(\gamma(G_x \cap G_y)) = \varphi(G_x \cap G_y) \ne 1$. Hence, $1 < \gamma(G_x \cap G_y) \le \gamma(G_x) \cap \gamma(G_y) \le B_{\gamma(x)} \cap B_{\gamma(y)}$. Write $\gamma(x) = \alpha'(i)^b$ and $\gamma(y) = \alpha'(i')^{b'}$ with $i, i' \in I_0$ and $b, b' \in B$. Then $B_i^b \cap B_{i'}^{b'} = B_{\alpha'(i)}^b \cap B_{\alpha'(i')}^{b'} = B_{\gamma(x)} \cap B_{\gamma(y)} \ne 1$. By Proposition 4.4, $i = i'$. Hence $\gamma(x)$ and $\gamma(y)$ are in the same B-orbit. Therefore, $\varphi(x)$ and $\varphi(y)$ are in the same A-orbit. The choice of φ gives $g \in G$ with $x^g = y$.

By the preceding paragraph, $B_{\gamma(x)} \cap B_{\gamma(x)}^{\gamma(g)} = B_{\gamma(x)} \cap B_{\gamma(x^g)} = B_{\gamma(x)} \cap B_{\gamma(y)} \neq 1$. Since $B_{\gamma(x)} = B_i^b$, Proposition 4.4 implies $\gamma(g) \in B_{\gamma(x)}$. Hence, $\varphi(g) \in A_{\varphi(x)}$. Since this relation holds for all φ with $\mathrm{Ker}(\varphi) \leq \mathrm{Ker}(\bar{\varphi})$, we have $g \in G_x$, as desired.

PROOF OF (b). Suppose $G_x \neq 1$. Consider $g \in G$ with $G_x^g = G_x$. By (a), there is an $a \in G_x$ with $x^{ga} = x$. Then $ga \in G_x$. Hence, $g \in G_x$.

PROOF OF (c). Suppose $G_x \neq G_y$ for some $x, y \in X$, then there is a $g \in G_x$ with $y = x^g$. Hence, by assumption, $y = x$. Thus, the forgetful map $\delta_{\mathbf{G}}$ is an étale continuous bijection of X onto $\mathcal{G} = \{G_x \mid x \in X\}$. By Corollary 1.4, $\mathcal{G}$ is étale Hausdorff. Since X is compact, $\delta_{\mathbf{G}}$ is an étale homeomorphism. It follows, $\mathbf{G}$ is proper. □

EXAMPLE 4.7. *Projective structures.*

(a) Projective group. Let G be a profinite group and X the empty space. Then $\mathbf{G} = (G, X, \ )$ is a projective proper group structure if and only if G is a projective group.

(b) Trivial stabilizers. Let G be an arbitrary profinite group. Put $X = G$. Then X is a profinite space and G acts continuously on X by multiplication from the right. In particular, $S_x = 1$ for each $x \in X$. For each $x \in X$ put $G_x = G$. Then $\mathbf{G} = (G, X, G_x)_{x \in X}$ is a projective group structure.

Indeed, let $(\varphi: \mathbf{G} \to \mathbf{A}, \alpha: \mathbf{B} \to \mathbf{A})$ be a finite embedding problem for $\mathbf{G}$. Let $i \in I$ and $j \in J$ be elements with $\varphi(1) = i$ and $\alpha(j) = i$. Then $G_1 = G$, so $\varphi(G) \leq A_i$. Also, $\alpha: B_j \to A_i$ is an isomorphism. Hence, $\gamma_{\mathrm{g}} = (\alpha|_{B_j})^{-1} \circ \varphi$ is a homomorphism from G to B satisfying $\alpha \circ \gamma_{\mathrm{g}} = \varphi$. Define $\gamma_{\mathrm{s}}: X \to J$ by $\gamma_{\mathrm{s}}(x) = j^{\gamma_{\mathrm{g}}(x)}$. Then $\gamma = (\gamma_{\mathrm{g}}, \gamma_{\mathrm{s}})$ is a solution of the embedding problem, as desired.

If G is nontrivial, then $1 \notin \mathrm{StrictClosure}\{G_x \mid x \in X\}$ but $\mathbf{G}$ is not proper. It follows that the assumption $S_x \neq G_x$ in Proposition 4.6(c) (which is violated in our example) is necessary.

(c) Free products of finitely many profinite groups. Let K be a finite set and K_0 a subset. For each $k \in K$ let G_k be a nontrivial profinite group. Suppose G_k is projective for each $k \in K \smallsetminus K_0$. Write $G = \mathop{\ast}_{k \in K} G_k$ for the free product of the G_k's. For each $k \in K$ the orbit $\mathcal{G}_k = \{G_k^g \mid g \in G\}$ of G_i under conjugation is a strictly closed subset of $\mathrm{Subgr}(G)$. Hence, $\mathcal{G} = \bigcup_{k \in K_0} \mathcal{G}_k$ is a strictly profinite subspace of $\mathrm{Subgr}(G)$, so strictly closed. In particular, $1 \notin \mathrm{StrictClosure}(\mathcal{G})$. By Proposition 4.4, $H \cap H' = 1$ for all distinct $H, H' \in \mathcal{G}$. It follows from Corollary 1.4 that $\mathcal{G}$ is étale Hausdorff.

Choose a homeomorphic copy X of $\mathcal{G}$ with the strict topology and a strict homeomorphism $\delta: X \to \mathcal{G}$. Since the strict topology of $\mathrm{Subgr}(G)$ is finer than its étale topology, δ is étale continuous. Since $\mathcal{G}$ is étale Hausdorff, δ is an étale homeomorphism. For each $x \in X$ let $G_x = \delta(x)$. By Proposition 4.4, each $H \in \mathcal{G}$ is its own normalizer in G. Thus, in the terminology of Section 2, $S_x = G_x$. Therefore, $\mathbf{G} = (G, X, G_x)_{x \in X}$ is a proper group structure.

We prove $\mathbf{G}$ is projective. To this end consider finite group structures $\mathbf{A} = (A, I, A_i)_{i \in I}$ and $\mathbf{B} = (B, J, B_j)_{j \in J}$, a cover $\alpha: \mathbf{B} \to \mathbf{A}$, and an epimorphism $\varphi: \mathbf{G} \to \mathbf{A}$. By Lemma 4.1 it suffices to find a morphism $\gamma: \mathbf{G} \to \mathbf{B}$ with $\gamma \circ \alpha = \varphi$.

Choose a map $\alpha': I \to J$ with $\alpha(\alpha'(i)) = i$ for each $i \in I$. Now consider $k \in K$. If $k \in K_0$, let x_k be the unique element of X with $\delta(x_k) = G_k$, $i = \varphi(k)$, and $j = \alpha'(i)$. Then $\alpha: B_j \to A_i$ is an isomorphism. Hence, $\gamma_k = (\alpha|_{B_j})^{-1} \circ (\varphi|_{G_k})$

is an epimorphism of G_k onto B_j satisfying $\alpha \circ \gamma_k = \varphi|_{G_k}$. If $k \in K \smallsetminus K_0$, then G_k is projective and we choose a homomorphism $\gamma_k: G_k \to B$ satisfying $\alpha \circ \gamma_k = \varphi|_{G_k}$. The basic property of free products gives a homomorphism $\gamma: G \to B$ whose restriction to each G_k is γ_k. In particular, $\alpha \circ \gamma = \varphi$. Together with the map $\gamma = \alpha' \circ \varphi$ from X to B, $\gamma: \mathbf{G} \to \mathbf{B}$ is a morphism satisfying $\alpha \circ \gamma = \varphi$, as desired.□

5. Special Covers

As in Lemma 4.5 we consider a group structure $\mathbf{G} = (G, X, G_x)_{x \in X}$ and an epimorphism of profinite groups $\pi: H \to G$. In contrast to Lemma 4.5, we do not assume that X has only finitely many G-orbits. Nor do we assume that X has a fundamental domain (beginning of Section 3). Nevertheless, we are able to extend $\pi: H \to G$ to a cover $\pi: \mathbf{H} \to \mathbf{G}$ in special cases described in Lemma 5.1 below. They occur three times in Galois-theoretic set-ups (in Lemma 14.2 and twice in Lemma 15.1).

LEMMA 5.1. *Let $\mathbf{G} = (G, X, G_x)_{x \in X}$ be a group structure and $(G_i, X_i)_{i \in I_0}$ a special partition of $\mathbf{G}$ (Definition 3.5). Let $\pi: H \to G$ be an epimorphism of profinite groups. For each $i \in I_0$ let H_i be a subgroup of H which π maps isomorphically onto G_i.*

Then H extends to a profinite group structure $\mathbf{H} = (H, Y, H_y)_{y \in Y}$ and π extends to a cover $\mathbf{H} \to \mathbf{G}$. Moreover, for each $i \in I_0$ there is a subspace Y_i of Y such that $\pi: Y_i \to X_i$ is a homeomorphism, $H_y \le H_i$ for each $y \in Y_i$, and $\bigcup_{i \in I_0} Y_i^H = Y$.

If, in addition, $\mathbf{G}$ is proper and

(1) $H_i^\kappa \cap H_i = 1$ for all $\kappa \in \mathrm{Ker}(\pi)$ with $\kappa \neq 1$ and each $i \in I_0$,

then $\mathbf{H}$ is proper.

PROOF. The proof has four parts.

PART A. *The space $\hat{Y}$.* Let $X' = \bigcup_{i \in I_0} X_i$. This is a profinite space and hence so is the product $\hat{Y} = X' \times H$. The group H acts continuously on $\hat{Y}$ by $(x,h)^\eta = (x, h\eta)$ and there is a continuous map $\hat{\pi}: \hat{Y} \to X$ defined by $\hat{\pi}(x,h) = x^{\pi(h)}$. Since $X = \bigcup_{i \in I_0} \bigcup_{\rho \in R_i} X_i^\rho$, this map is surjective.

For each $y = (x,h) \in \hat{Y}$ define a subgroup H_y of H in the following way. There is a unique $i \in I_0$ with $x \in X_i$. Then $G_x \le G_i$. Let H_x be the unique subgroup of H_i satisfying $\pi(H_x) = G_x$. Put $H_y = H_x^h$. Then

(2a) $\hat{\pi}(y^\eta) = \hat{\pi}(y)^{\pi(\eta)}$ for all $y \in \hat{Y}$ and $\eta \in H$,

(2b) $H_y^\eta = H_{y^\eta}$ for all $y \in \hat{Y}$ and $\eta \in H$, and

(2c) $\pi: H_y \to G_{\hat{\pi}(y)}$ is an isomorphism, $y \in \hat{Y}$.

CLAIM A1. *The map $\hat{\delta}_n: \hat{Y} \to \mathrm{Subgr}(H)$ defined by $\hat{\delta}(y) = H_y$ is étale continuous.* It suffices to prove that the map $X_i \to \mathrm{Subgr}(H)$ defined by $x \mapsto H_x$ is étale continuous. By Remark 2.6 we have to prove that the corresponding map $X_i \to \mathrm{Subgr}(H_i)$ is étale continuous. Now, by assumption, the map $X \to \mathrm{Subgr}(G)$ given by $x \mapsto G_x$ is étale continuous. Hence, by Remark 2.6, the corresponding map $X_i \to \mathrm{Subgr}(G_i)$ is continuous. Since G_i is isomorphic to H_i, we get our claim.

EQUIVALENCE RELATION. Define an equivalence relation $\equiv$ on $\hat{Y}$ as follows. Let $(x_1, h_1) \equiv (x_2, h_2)$ if there is a (unique) $i \in I_0$ with $x_1, x_2 \in X_i$, $H_i h_1 = H_i h_2$, and $x_1^{\pi(h_1)} = x_2^{\pi(h_2)}$. This relation satisfies the following rules:

(3a) If $y_1 \equiv y_2$, then $\hat{\pi}(y_1) = \hat{\pi}(y_2)$.

(3b) If $y_1 \equiv y_2$, then $H_{y_1} = H_{y_2}$.

Indeed, both H_{x_1} and $H_{x_2}^{h_2 h_1^{-1}}$ are contained in H_i for some $i \in I$ and $\pi(H_{x_1}) = \pi(H_{x_2}^{h_2 h_1^{-1}})$. Hence, $H_{x_1} = H_{x_2}^{h_2 h_1^{-1}}$, so $H_{y_1} = H_{y_2}$.

(3c) If $y_1 \equiv y_2$ and $\eta \in H$, then $y_1^\eta \equiv y_2^\eta$.

Let $K = \mathrm{Ker}(\pi)$.

CLAIM A2. *$\hat{\pi}(x_1, h_1) = \hat{\pi}(x_2, h_2)$ if and only if there is a $k \in K$ with $(x_2, h_2) \equiv (x_1, h_1 k)$.*

Indeed, let $i \in I_0$ with $x_1, x_2 \in X_i$. If $\hat{\pi}(x_1, h_1) = \hat{\pi}(x_2, h_2)$, then $x_1^{\pi(h_1 h_2^{-1})} = x_2$. Hence, by (2d) and (2f) of Section 3, $\pi(h_1 h_2^{-1}) \in G_i = \pi(H_i)$. Therefore, there is a $k_0 \in K$ with $h_1 h_2^{-1} k_0 \in H_i$. Then $k = h_2^{-1} k_0 h_2 \in K$, $H_i h_1 k = H_i h_2$, and $x_1^{\pi(h_1 k)} = x_1^{\pi(h_1)} = x_2^{\pi(h_2)}$. Consequently $(x_2, h_2) \equiv (x_1, h_1 k)$.

Conversely, if $(x_2, h_2) \equiv (x_1, h_1 k)$, then $x_2^{\pi(h_2)} = x_1^{\pi(h_1 k)} = x_1^{\pi(h_1)}$, so $\hat{\pi}(x_2, h_2) = \hat{\pi}(x_1, h_1)$.

PART B. *The quotient space Y.* Let Y be the quotient space of $\hat{Y}$ modulo $\equiv$. By (3a), $\hat{\pi}: \hat{Y} \to X$ induces a continuous surjection $\pi: Y \to X$. By (3b), $\hat{\delta}_H: \hat{Y} \to \mathrm{Subgr}(H)$ induces a étale continuous map $\delta_H: Y \to \mathrm{Subgr}(H)$. By (3c), the H-action on $\hat{Y}$ induces a continuous action of H on Y. By (2),

(4a) $\pi(y^\eta) = \pi(y)^{\pi(\eta)}$ for all $y \in Y$ and $\eta \in H$,

(4b) $H_y^\eta = H_{y^\eta}$ for all $y \in Y$ and $\eta \in H$, and

(4c) $\pi: H_y \to G_{\pi(y)}$ is an isomorphism, for each $y \in Y$.

Finally, by Claim A2,

(5) $\pi(y_1) = \pi(y_2)$ if and only if there is a $k \in K$ with $y_2 = y_1^k$.

CLAIM B1. *Y is a profinite space.* Indeed, $\hat{Y}$ is compact, hence so is Y. Consider inequivalent $y_1, y_2 \in \hat{Y}$. It suffices to produce an open-closed neighborhood U of y_1 which is closed under $\equiv$ and does not contain y_2.

If $\hat{\pi}(y_1) \neq \hat{\pi}(y_2)$, we choose an open-closed neighborhood V of $\hat{\pi}(y_1)$ in X which does not contain $\hat{\pi}(y_2)$. Then $U = \hat{\pi}^{-1}(V)$ has the required property.

If $\hat{\pi}(y_1) = \hat{\pi}(y_2)$, we use Claim A2 to replace y_2 by an equivalent element of $\hat{Y}$ to assume that $y_1 = (x_1, h_1)$, $y_2 = (x_1, h_1 k)$, where $1 \neq k \in K$. Let $i \in I_0$ such that $x_1 \in X_i$. Then $H_i \cap K = 1$, so $h_1 k h_1^{-1} \notin H_i$. There is an open subgroup H_i' which contains H_i and $h_1 k h_1^{-1} \notin H_i'$. Let $U = X_i \times H_i' h_1$. Then $(x_1, h_1) \in U$ but $h_1 k \notin H_i' h_1$, so $(x_1, h_1 k) \notin U$. Clearly U is an open-closed subset of $\hat{Y}$ closed under $\equiv$.

CLAIM B2. *The stabilizer S_y of each $y \in Y$ is contained in H_y.*

Indeed, let y be represented by $(x, h) \in \hat{Y}$. Let $i \in I_0$ with $x \in X_i$. Let $\eta \in H$. Then

$$\begin{aligned} y^\eta = y &\Longrightarrow (x, h)^\eta \equiv (x, h) \\ &\Longrightarrow (x, h\eta) \equiv (x, h) \\ &\Longrightarrow H_i h\eta = H_i h \text{ and } x^{\pi(h\eta)} = x^{\pi(h)} \\ &\Longrightarrow \eta \in H_i^h \text{ and } \pi(\eta) \in S_{x^{\pi(h)}}. \end{aligned}$$

Hence, $S_y \leq H_i^\eta$ and $\pi(S_y) \leq S_{x^{\pi(h)}} \leq G_{x^{\pi(h)}} \leq G_i^{\pi(h)}$. In addition, π maps H_i^h isomorphically onto $G_i^{\pi(h)}$ and $\pi(H_y) = G_{x^{\pi(h)}}$. Therefore, $S_y \leq H_y$, as claimed.

Claim B2 completes the proof that $\mathbf{H} = (H, Y, H_y)_{y \in Y}$ is a group structure and $\pi: \mathbf{H} \to \mathbf{G}$ is a cover.

PART C. *The spaces* Y_i. For each $i \in I_0$ let Y_i be the image of $X_i \times 1$ in Y. Then, π maps Y_i homeomorphically onto X_i. By definition, $H_y \le H_i$ for each $y \in Y_i$. By the assumption on X we have $X = \bigcup_{i \in I_0} X_i^G$. Since $\pi: \mathbf{H} \to \mathbf{G}$ is a cover and $\pi(Y_i) = X_i$, we have $Y = \bigcup_{i \in I_0} Y_i^H$.

PART D. $\mathbf{H}$ *is proper under the assumption that* $\mathbf{G}$ *is proper and (1) holds.* Indeed, let $\mathcal{H} = \{H_y \mid y \in Y\}$ and $\mathcal{G} = \{G_x \mid x \in X\}$. Since $\pi(\mathcal{H}) = \mathcal{G}$, we have $\pi(\mathrm{StrictClosure}(\mathcal{H})) \subseteq \mathrm{StrictClosure}(\mathcal{G})$. Since 1 is not in $\mathrm{StrictClosure}(\mathcal{G})$, it is not in $\mathrm{StrictClosure}(\mathcal{H})$.

Let $y_1, y_2 \in Y$ be distinct. We prove that H_{y_1}, H_{y_2} are distinct and can be separated in the étale topology of $\mathrm{Subgr}(H)$.

First suppose $\pi(y_1) \ne \pi(y_2)$. Then $G_{\pi(y_1)} \ne G_{\pi(y_2)}$. Since $\mathcal{G}$ is étale profinite, there are open subgroups E_1, E_2 of G with $\pi(H_{y_i}) = G_{\pi(y_i)} \le E_i$, $i = 1, 2$, and $\mathcal{G} \cap \mathrm{Subgr}(E_1) \cap \mathrm{Subgr}(E_2) = \emptyset$. Then $F_1 = \pi^{-1}(E_1)$ and $F_2 = \pi^{-1}(E_2)$ are open subgroups of H, $H_{y_1} \le F_1$, $H_{y_2} \le F_2$, and $\mathcal{H} \cap \mathrm{Subgr}(F_1) \cap \mathrm{Subgr}(F_2) = \emptyset$.

Now suppose $\pi(y_1) = \pi(y_2)$. Since π is a cover, there is a $\kappa \in K$ with $y_2 = y_1^\kappa$. Since $y_1 \ne y_2$, we have $\kappa \ne 1$. Let y_1 be represented by $(x, h) \in \hat{Y}$, with $x \in X_i$, where $i \in I_0$, and $h \in H$. Then $H_{y_1} = H_x^h \le H_i^h$ and $H_{y_2} = H_x^{h\kappa} \le H_i^{h\kappa}$. By (1), $H_i^{h\kappa h^{-1}} \cap H_i = 1$, that is, $H_i^{h\kappa} \cap H_i^h = 1$. Hence, $H_{y_1} \cap H_{y_2} = 1$. By Corollary 1.4(a), H_{y_1} and H_{y_2} can be separated by the étale topology of $\mathcal{H}$.

It follows that $\mathcal{H}$ is étale Hausdorff and the étale continuous map $\delta_H: Y \to \mathrm{Subgr}(H)$ is bijective. By Claim B1, Y is compact. Hence, δ_H is a homeomorphism. Consequently, $\mathbf{H}$ is proper. □

6. Unirationally Closed Fields

Galois correspondence naturally translates group structures $\mathbf{G} = (G, X, G_x)_{x \in X}$ with $G = \mathrm{Gal}(K)$ and K a field to "field structures" $\mathbf{K} = (K, X, K_x)_{x \in X}$ with $\mathrm{Gal}(K_x) = G_x$ for each $x \in X$. We give an arithmetically geometric criterion for $\mathbf{G}$ to be projective. It generalizes Ax's theorem saying that $\mathrm{Gal}(K)$ is projective if K is PAC. The standard proof of Ax's result [FrJ, p. 207] actually uses only the existence of K-rational points on varieties over K which become unirational over a finite extension of K. Our criterion has the same nature amended with a local-global flavor.

Let K be a field. Denote the set of all algebraic (resp. separable algebraic) extensions of K by $\mathrm{AlgExt}(K)$ (resp. $\mathrm{SepAlgExt}(K)$). Galois theory puts $\mathrm{SepAlgExt}(K)$ in a bijective order-reversing correspondence with $\mathrm{Subgr}(\mathrm{Gal}(K))$. It equips $\mathrm{SepAlgExt}(K)$ with two natural topologies, the **strict topology** and the **étale topology**. A basic étale open subset of $\mathrm{SepAlgExt}(K)$ is $\mathrm{SepAlgExt}(L)$, where L is a finite extension of K. Thus, $\mathrm{SepAlgExt}(K)$ is not étale Hausdorff unless $K = K_s$. A basic strictly open subset of $\mathrm{SepAlgExt}(K)$ is

$$\{K' \in \mathrm{SepAlgExt}(K) \mid L \cap K' = L_0\},$$

where L_0 is a finite separable extension of K and L is a finite Galois extension of K containing L_0. $\mathrm{SepAlgExt}(K)$ is a profinite space under the strict topology. Denote the strict closure of a subset $\mathcal{X}$ of $\mathrm{SepAlgExt}(K)$ by $\mathrm{StrictClosure}(\mathcal{X})$.

A **field structure** is a triple $\mathbf{K} = (K, X, \delta)$ consisting of a field K, a profinite space X, an étale continuous map $\delta: X \to \mathrm{SepAlgExt}(K)$, and an étale continuous action (from the right) of $\mathrm{Gal}(K)$ on X satisfying the following condition:

(1a) For each $x \in X$ put $K_x = \delta(x)$. Then $K_{x^\sigma} = K_x^\sigma$ for all $x \in X$ and $\sigma \in \mathrm{Gal}(K)$.

(1b) $x \in X$, $\sigma \in \mathrm{Gal}(K)$, and $x^\sigma = x$ imply $\sigma \in \mathrm{Gal}(K_x)$.

As with group structures, we usually write $\mathbf{K}$ as $(K, X, K_x)_{x \in X}$. The **absolute Galois group structure** associated with $\mathbf{K}$ is

$$\mathrm{Gal}(\mathbf{K}) = (\mathrm{Gal}(K), X, \mathrm{Gal}(K_x))_{x \in X}.$$

Conversely, to each absolute Galois group structure $\mathbf{G} = (\mathrm{Gal}(K), X, G_x)_{x \in X}$ we associate a field structure $\mathbf{K} = (K, X, K_x)_{x \in X}$, where K_x is the fixed field of G_x in K_s. Then $\mathrm{Gal}(\mathbf{K}) = \mathbf{G}$. We use the correspondence between field structures and absolute Galois group structures to translate the terminology and results obtained so far from group structures to field structures.

DEFINITION 6.1. A **unirational arithmetical problem** for a field structure $\mathbf{K} = (K, X, K_x)_{x \in X}$ is a data

(2) $$\Phi = (V, X_i, L_i, \pi_i: U_i \to V \times_K L_i)_{i \in I_0}$$

satisfying these conditions:

(3a) $(\mathrm{Gal}(L_i), X_i)_{i \in I_0}$ is a special partition of $\mathrm{Gal}(\mathbf{K})$ (Definition 3.5).

(3b) V is a smooth absolutely irreducible affine variety defined over K.

(3c) U_i is a smooth absolutely irreducible affine variety over L_i birationally equivalent to $\mathbb{A}_{L_i}^{\dim(V)}$.

(3d) $\pi_i: U_i \to V \times_K L_i$ is an étale morphism.

Let $X' = \bigcup_{i \in I_0} X_i$. A **solution** of Φ is an "extended point" $(\mathbf{a}, \mathbf{b}_x)_{x \in X'}$ with $\mathbf{a} \in V(K)$, $\mathbf{b}_x \in U_i(K_x)$, and $\pi_i(\mathbf{b}_x) = \mathbf{a}$ for each $i \in I_0$ and all $x \in X_i$. Call $\mathbf{K}$ **unirationally closed** if each unirational arithmetical problem for $\mathbf{K}$ has a solution. □

LEMMA 6.2 [HJK, LEMMA 3.1]. *Let L/K be a finite Galois extension and $\psi: B \to \mathrm{Gal}(L/K)$ an epimorphism of finite groups. Then there exists a finitely generated regular extension E of K and a finite Galois extension F of E containing L such that $B = \mathrm{Gal}(F/E)$ and ψ is the restriction $\mathrm{res}_{F/L}: \mathrm{Gal}(F/E) \to \mathrm{Gal}(L/K)$.*

Moreover, let $K \subseteq L_0 \subseteq L$ and $E \subseteq F_0 \subseteq F$ be fields with $L_0 \subseteq F_0$. Suppose $\psi: \mathrm{Gal}(F/F_0) \to \mathrm{Gal}(L/L_0)$ is an isomorphism. Then F_0 is a purely transcendental extension of L_0 of transcendence degree $|B|$.

PROOF. Let x^β, $\beta \in B$, be algebraically independent elements over K. Define a faithful action of B on $F = L(x^\beta \mid \beta \in B)$ by $(x^\beta)^{\beta'} = x^{\beta\beta'}$ and $a^{\beta'} = a^{\psi(\beta')}$ for $a \in L$. Denote the fixed field of B in F by E. Then F/K is a finitely generated separable extension. By [Lan, p. 64, Prop. 6], E/K is also a finitely generated separable extension. Also, $\mathrm{res}: \mathrm{Gal}(F/E) \to \mathrm{Gal}(L/K)$ coincides with $\psi: B \to \mathrm{Gal}(L/K)$. Hence, $E \cap \tilde{K} = E \cap F \cap \tilde{K} = E \cap L = K$. Therefore, E/K is regular.

Now let L_0 and F_0 as in the second paragraph of the lemma. Put $B_0 = \mathrm{Gal}(F/F_0)$. Choose a set of representatives R for the left cosets of B modulo B_0. Let $w_1, \ldots, w_m$ be a basis for L/L_0. By assumption, $m = |B_0|$. Consider $\rho \in R$.

Put

$$t_{\rho j} = \sum_{\beta \in B_0} w_j^\beta x^{\rho\beta}, \qquad j = 1, \dots, m.$$

Since $\det(w_j^\beta) \neq 0$, each $x^{\rho\beta}$ is a linear combination of $t_{\rho j}$ with coefficients in L. Put $\mathbf{t} = (t_{\rho j} \mid \rho \in R,\ j = 1, \dots, m)$, $\mathbf{x} = (x^\beta \mid \beta \in B)$, and $n = |B|$. Both tuples contain exactly n elements and $L(\mathbf{t}) = L(\mathbf{x}) = F$. So, $L_0(\mathbf{t})$ is a purely transcendental extension of L_0.

Each $t_{\rho j}$ is fixed by B_0. Hence, $L_0(\mathbf{t}) \subseteq F_0$. Moreover, $m = [L : L_0] = [L(\mathbf{t}) : L_0(\mathbf{t})] \geq [F : F_0] = |B_0| = m$. Consequently $F_0 = L_0(\mathbf{t})$ and F_0/L_0 is purely transcendental. □

LEMMA 6.3. *Let $\mathbf{G} = (\mathbf{G}, X, G_x)_{x \in X}$ and $\mathbf{A} = (\mathbf{A}, I, A_i)_{i \in I}$ be group structures and $\varphi\colon \mathbf{G} \to \mathbf{A}$ an epimorphism. Suppose $S_x = G_x$ for each $x \in X$. Then $S_i = A_i$ for each $i \in I$.*

PROOF. By assumption, $S_i \leq A_i$. Conversely, let $a \in A_i$. By assumption, there is $x \in X$ with $\varphi(x) = i$ and $\varphi(G_x) = A_i$. Choose $g \in G_x$ with $\varphi(g) = a$. Then $i^a = \varphi(x)^{\varphi(g)} = \varphi(x^g) = \varphi(x) = i$. Thus, $a \in S_i$. □

PROPOSITION 6.4. *Let $\mathbf{K} = (K, X, K_x)_{x \in X}$ be a unirationally closed field structure. Suppose $S_x = \mathrm{Gal}(K_x)$ for each $x \in X$. Then $\mathrm{Gal}(\mathbf{K})$ is a projective group structure.*

PROOF. Let $(\varphi\colon \mathrm{Gal}(\mathbf{K}) \to \mathbf{A},\ \alpha\colon \mathbf{B} \to \mathbf{A})$ be a finite embedding problem for $\mathrm{Gal}(\mathbf{K})$. Thus $\mathbf{B} = (B, J, B_j)_{j \in J}$ and $\mathbf{A} = (A, I, A_i)_{i \in I}$ are finite group structures and α is a cover. By Lemma 4.1, we may assume φ is an epimorphism. By assumption, I is discrete and the map $\varphi\colon X \to I$ is continuous. Hence, $X_i = \{x \in X \mid \varphi(x) = i\}$ is an open-closed subset of X, $i \in I$ and $X = \bigcup_{i \in I} X_i$. Moreover, $X_i^\sigma = X_{i^{\varphi(\sigma)}}$ for all $i \in I$ and $\sigma \in \mathrm{Gal}(K)$.

Choose a set of representatives I_0 for the A-orbits of I and for each $i \in I_0$ choose $j(i) \in J$ with $\alpha(j(i)) = i$.

The rest of the proof has six parts.

PART A. *Replacing* **A** *and* **B** *by Galois structures.* Replace A by $\mathrm{Gal}(L/K)$, where L is a finite Galois extension of K, to assume that $\varphi\colon \mathrm{Gal}(K) \to \mathrm{Gal}(L/K)$ is $\mathrm{res}_{K_s/L}$. Denote the fixed field of A_i in L by L_i. By assumption, $S_x = \mathrm{Gal}(K_x)$ for each $x \in X$. Hence, by Lemma 6.3, $\mathrm{Gal}(L/L_i) = \{\sigma \in \mathrm{Gal}(L/K) \mid i^\sigma = i\}$. Therefore,

$$\mathrm{Gal}(L_i) = \{\sigma \in \mathrm{Gal}(K) \mid X_i^\sigma = X_i\}. \tag{4}$$

and $(\mathrm{Gal}(L_i), X_i, \mathrm{Gal}(K_x))_{x \in X_i}$ is a group structure.

Lemma 6.2 gives a finitely generated regular extension E of K and a finite Galois extension F of E containing L and allows us to replace B by $\mathrm{Gal}(F/E)$ and $\alpha\colon B \to \mathrm{Gal}(L/K)$ by $\mathrm{res}_{F/L}\colon \mathrm{Gal}(F/E) \to \mathrm{Gal}(L/K)$. For each $i \in I_0$ denote the fixed field of $B_{j(i)}$ in F by F_i. Since α is a cover, $\mathrm{res}_{F/L}\colon \mathrm{Gal}(F/F_i) \to \mathrm{Gal}(L/L_i)$ is an isomorphism. Hence, by Lemma 6.2, F_i is a purely transcendental extension of L_i of transcendence degree $r = [F : E]$.

Since $\varphi\colon \mathrm{Gal}(\mathbf{K}) \to \mathbf{A}$ is a morphism, $\mathrm{res}_{K_s/L}(\mathrm{Gal}(K_x)) \le \mathrm{Gal}(L/L_i)$ for all $x \in X_i$. Hence, $L_i \le K_x$.

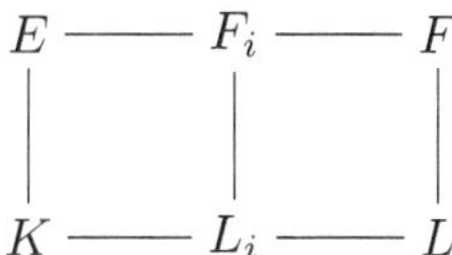

PART B. *Setting up a unirational arithmetical problem.* Choose $y_1, \dots, y_n \in E$, $z_i \in F_i$, and $\tilde{z} \in F$ satisfying this:

(5a) $E = K(\mathbf{y})$ and $V = \mathrm{Spec}(K[\mathbf{y}])$ is a smooth affine absolutely irreducible subvariety of $\mathbb{A}_K^n$ define over K with generic point $\mathbf{y}$ and $\dim(V) = r$.

(5b) For each $i \in I_0$ the following holds: $F_i = L_i(\mathbf{y}, z_i)$ and $U_i = \mathrm{Spec}(L_i[\mathbf{y}, z_i])$ is a smooth Zariski closed affine subvariety of $\mathbb{A}_{L_i}^{n+1}$ define over L_i and birationally equivalent to $\mathbb{A}_{L_i}^r$ with generic point $(\mathbf{y}, z_i)$.

(5c) z_i is integral over $L_i[\mathbf{y}]$ and the discriminant of $\mathrm{irr}(z_i, L_i(\mathbf{y}))$ is a unit of $L_i[\mathbf{y}]$. Hence, $L_i[\mathbf{y}, z_i]/L_i[\mathbf{y}]$ is a ring cover in the terminology of [FrJ, Definition 6.1.3]. Thus, projection on the first n coordinates is an étale morphism $\pi_i\colon U_i \to V \times_K L_i$.

(5d) $F = K(\mathbf{y}, \tilde{z})$ and $L[\mathbf{y}, \tilde{z}]/L[\mathbf{y}]$ is a ring cover.

Then, (2) is a unirational arithmetical problem for $\mathbf{K}$ satisfying Condition (3).

PART C. *A solution of a unirational arithmetical problem.* Since $\mathbf{K}$ is unirationally closed, Problem (2) has a solution. Thus, there are $\mathbf{a} \in V(K)$ and $\mathbf{b}_x = (\mathbf{a}, c_x) \in U_i(K_x)$ for each $i \in I_0$ and all $x \in X_i$.

Let $i \in I_0$ and $x \in X_i$. Then $\mathrm{Gal}(L_i(c_x))$ is an open subgroup of $\mathrm{Gal}(L_i)$ which contains $\mathrm{Gal}(K_x)$. Also, $W_x = \{x' \in X_i \mid L_i(c_x) \subseteq K_{x'}\}$ is an étale open subset of X_i which contains $x^{\mathrm{Gal}(L_i(c_x))}$. Lemma 3.6 with $\mathrm{Gal}(L_i), X_i, X_i, x, \mathrm{Gal}(L_i(c_x)), W_x$ respectively replacing G, X, Y, y, G'_y, V_y, gives

(6a) a finite set Λ_i, and

(6b) for each $l \in \Lambda_i$ an open-closed subset X_{il} of X_i, an element $x_{il} \in X_{il}$, and a finite subset T_{il} of $\mathrm{Gal}(L_i)$,

such that $(\mathrm{Gal}(L_i(c_{il})), T_{il}, X_{il})_{l \in \Lambda_i}$ is a special partition of

$$(\mathrm{Gal}(L_i), X_i, \mathrm{Gal}(K_x))_{x \in X_i},$$

where $c_{il} = c_{x_{il}}$. Thus,

(7a) $\mathrm{Gal}(L_i(c_{il})) = \{\sigma \in \mathrm{Gal}(K) \mid X_{il}^\sigma = X_{il}\}$ for each $l \in \Lambda_i$,

(7b) $\mathrm{Gal}(L_i) = \bigcup_{\tau \in T_{il}} \mathrm{Gal}(L_i(c_{il}))\tau$, and

(7c) $X_i = \bigcup_{l \in \Lambda_i} \bigcup_{\tau \in T_{il}} X_{il}^\tau$.

PART D. *A homomorphism* $\gamma\colon G \to B$. Since $\mathbf{a}$ is simple on V, there is a K-place $\rho\colon E \to K \cup \{\infty\}$ with $\rho(\mathbf{y}) = \mathbf{a}$ [JaR, Cor. A2]. Extend ρ to an L-place $\rho\colon F \to \tilde{K} \cup \{\infty\}$. Let $\bar{F}$ be the residue field of ρ. By (5d), $\bar{F}$ is a finite Galois extension of K containing L [FrJ, Lemma 6.1.4]. Moreover, there is an embedding $\rho^*\colon \mathrm{Gal}(\bar{F}/K) \to \mathrm{Gal}(F/E)$ with $\rho(\rho^*(\sigma)u) = \sigma(\rho(u))$ for all $\sigma \in \mathrm{Gal}(\bar{F}/K)$ and $u \in F$ with $\varphi(u) \ne \infty$ [FrJ, Lemma 6.1.4]. Then $\gamma = \rho^* \circ \mathrm{res}_{K_s/\bar{F}}$ is a homomorphism from $\mathrm{Gal}(K)$ to $\mathrm{Gal}(F/E)$ with $\mathrm{res}_{F/L} \circ \gamma = \mathrm{res}_{K_s/L}$.

PART E. *A continuous map* $\gamma\colon X \to J$. Let $i \in I_0$ and $l \in \Lambda_i$. By (5b), $(\mathbf{a}, c_{il})$ is simple on U_i. So, there is an L_i-place $\rho_{il}\colon F_i \to L_i(c_{il}) \cup \{\infty\}$ with $\rho_{il}(\mathbf{y}, z_i) = (\mathbf{a}, c_{il})$ [JaR, Cor. A2]. Extend it to an L-place $\rho_{il}\colon F \to \tilde{K} \cup \{\infty\}$.

Since $\rho_{il}|_{EL} = \rho|_{EL}$, there is a $\sigma_{il} \in \mathrm{Gal}(F/EL)$ and $\rho_{il} = \rho \circ \sigma_{il}^{-1}$. Define γ on X_{il} as the constant map: $\gamma(x) = j(i)^{\sigma_{il}}$ for all $x \in X_{il}$. Then $\rho(F_i^{\sigma_{il}}) = \rho \circ \sigma_{il}^{-1}(F_i) = \rho_{il}(F_i) \subseteq L_i(c_{il}) \cup \{\infty\} \subseteq K_x \cup \{\infty\}$. This implies

(8) $\gamma(\mathrm{Gal}(K_x)) \le \gamma(\mathrm{Gal}(L_i(c_{il}))) \le \mathrm{Gal}(F/F_i)^{\sigma_{il}} = B_{j(i)^{\sigma_{il}}} = B_{\gamma(x)}, \quad x \in X_{il}.$

Also, $\alpha(\gamma(x)) = \alpha(j(i)^{\sigma_{il}}) = \alpha(j(i)) = i = \varphi(x)$.

Let $X' = \bigcup_{i\in I_0} \bigcup_{l \in \Lambda_i} X_{il}$. Then $X = \bigcup_{\sigma \in \mathrm{Gal}(K)} (X')^\sigma$. Since each X_{il} is open, $\gamma\colon X' \to J$ is well defined and continuous. Extend $\gamma\colon X' \to J$ to X by $\gamma(x^\sigma) = \gamma(x)^{\gamma(\sigma)}$ for each $\sigma \in \mathrm{Gal}(K)$. To prove this is a good definition, we have to show that if $x, y \in X'$, $\sigma_1, \sigma_2 \in \mathrm{Gal}(K)$, and $x^{\sigma_1} = y^{\sigma_2}$, then $\gamma(x)^{\gamma(\sigma_1)} = \gamma(y)^{\gamma(\sigma_2)}$. In other words, with $\sigma = \sigma_1\sigma_2^{-1}$, we have to prove that

(9) $x \in X'$, $\sigma \in \mathrm{Gal}(K)$, and $x^\sigma \in X'$ imply $\gamma(x)^{\gamma(\sigma)} = \gamma(x^\sigma)$.

Indeed, there are $i, i' \in I_0$, $l \in \Lambda_i$, and $l' \in \Lambda_{i'}$ with $x \in X_{il}$ and $x^\sigma \in X_{i'l'}$. By (6b), $x \in X_i$ and $x^\sigma \in X_{i'}$. Hence, $x^\sigma \in X_i^\sigma \cap X_{i'} = X_{i^\sigma} \cap X_{i'}$. Therefore, $i^\sigma = i'$. By the choice of I_0, this implies $i = i'$. Hence $x^\sigma \in X_{i^\sigma} \cap X_i$, so $X_i^\sigma = X_i$. By (4), $\sigma \in \mathrm{Gal}(L_i)$. It follows, $x \in X_{il}$ and $x^\sigma \in X_{il'}$, so $x^\sigma \in X_{il}^\sigma \cap X_{il'}$. By (7b), $\sigma = \sigma'\tau$ with $\sigma' \in \mathrm{Gal}(L_i(c_{il}))$ and $\tau \in T_{il}$. Then, by (7c) and (7a), $X_{il'} = X_{il}^{\sigma'\tau} = X_{il}^\tau$. Hence, by (7c), $\tau \in \mathrm{Gal}(L_i(c_{il}))$, so also $\sigma \in \mathrm{Gal}(L_i(c_{il}))$. By (7c), $X_{il'} = X_{il}$. We have therefore proved both x and x^σ belong to X_{il}. By definition, $\gamma(x) = j(i)^{\sigma_{il}} = \gamma(x^\sigma)$. By (8), $\gamma(\sigma) \in \gamma(\mathrm{Gal}(L_i(c_{il}))) \le B_{\gamma(x)}$. Hence, $\varphi(\sigma) = \alpha(\gamma(\sigma)) \in A_{\varphi(x)}$. By assumption, $S_x = \mathrm{Gal}(K_x)$ for each $x \in X$. By Lemma 6.3, $\alpha(\gamma(\sigma)) \in S_{\varphi(x)}$. Hence, by Lemma 2.2, $\gamma(x)^{\gamma(\sigma)} = \gamma(x)$. Therefore $\gamma(x^\sigma) = \gamma(x) = \gamma(x)^{\gamma(\sigma)}$, as claimed.

PART F. *Conclusion of the proof.* By (9), $\gamma(x)^{\gamma(\sigma)} = \gamma(x^\sigma)$ for all $x \in X$ and $\sigma \in \mathrm{Gal}(K)$. Hence, by (8), $\gamma(\mathrm{Gal}(K_{x^\sigma})) \le B_{\gamma(x^\sigma)}$ for all $x \in X$. Therefore, $\gamma\colon \mathrm{Gal}(\mathbf{K}) \to \mathbf{B}$ is a morphism. Finally, $\alpha \circ \gamma = \varphi$ on $\mathrm{Gal}(K)$ and on X', hence on X. Thus, γ solves the embedding problem we posed for $\mathrm{Gal}(\mathbf{K})$. It follows, $\mathrm{Gal}(\mathbf{K})$ is projective. □

7. Valued Fields

The results of this section are well known, although there is some novelty in the presentation[1]. We begin with a brief review of inertia and ramification groups.

Denote the residue field of a valued field (F, v) by $\bar{F}$. For each $x \in F$ with $v(x) \ge 0$ let $\bar{x}$ be the residue of x in $\bar{F}$. Finally, let F_{ins} be the maximal purely inseparable extension of F.

Consider a Galois extension $(N, v)/(F, v)$ of Henselian fields. Then $\bar{N}/\bar{F}$ is a normal extension. For each $\sigma \in \mathrm{Gal}(N/F)$ define $\bar{\sigma} \in \mathrm{Aut}(\bar{N}/\bar{F})$ by this rule: $\bar{\sigma}\bar{x} = \overline{\sigma x}$ for $x \in N$ with $v(x) \ge 0$. The map $\sigma \mapsto \bar{\sigma}$ is an epimorphism $\rho\colon \mathrm{Gal}(N/F) \to \mathrm{Aut}(\bar{N}/\bar{F})$ [End, Thm. 19.6]. Its kernel is the **inertia group**:

$$G_0(N/F) = \{\sigma \in \mathrm{Gal}(N/F) \mid v(\sigma x - x) > 0 \text{ for each } x \in N \text{ with } v(x) \ge 0\}.$$

Denote the fixed field in N of $G_0(N/F)$ by N_0. Then $\bar{N}_0$ is the maximal separable extension of $\bar{F}$ in $\bar{N}$ [End, Thm. 19.12]. Hence, $\bar{N}_0/\bar{F}$ is Galois and there is a short exact sequence

$$(1) \qquad 1 \longrightarrow \mathrm{Gal}(N/N_0) \longrightarrow \mathrm{Gal}(N/F) \xrightarrow{\ \rho\ } \mathrm{Gal}(\bar{N}_0/\bar{F}) \longrightarrow 1.$$

[1]This section is a rewrite of the unpublished paper [HJK, Sec. 2].

Here we have identified each $\bar{\sigma} \in \mathrm{Aut}(\bar{N}/\bar{F})$ with its restriction to $\bar{N}_0$. In addition $v(N_0^\times) = v(F^\times)$ [End, Cor. 19.14]. Hence, N_0/F is an unramified extension.

The **ramification group** of $\mathrm{Gal}(N/F)$ is

$$G_1(N/F) = \{\sigma \in \mathrm{Gal}(N/F) \mid v\Big(\frac{\sigma x}{x} - 1\Big) > 0 \text{ for each } x \in N^\times\}.$$

It is a normal subgroup of $\mathrm{Gal}(N/F)$ which is contained in $G_0(N/F)$ [End, (20.8)]. Denote the fixed field of $G_1(N/F)$ in N by N_1. When $p = \mathrm{char}(\bar{F}) > 0$, $\mathrm{Gal}(N/N_1)$ is the unique p-Sylow subgroup of $\mathrm{Gal}(N/N_0)$ [End, Thm. 20.18]. When $\mathrm{char}(\bar{F}) = 0$, $\mathrm{Gal}(N/N_1)$ is trivial. So, in both cases, $\mathrm{char}(\bar{F})$ does not divide $[N_1 : N_0]$.

Now suppose $N = F_s$. Then $F_u = N_0$ is the **inertia field** and $F_r = N_1$ is the **ramification field** of F. In this case (1) becomes the short exact sequence

$$(2) \qquad 1 \longrightarrow \mathrm{Gal}(F_u) \longrightarrow \mathrm{Gal}(F) \xrightarrow{\ \rho\ } \mathrm{Gal}(\bar{F}) \longrightarrow 1.$$

Also, $F \subseteq F_u \subseteq F_r \subseteq F_s$, F_u/F and F_r/F are Galois extensions, $\mathrm{char}(\bar{F}) \nmid [F_r : F_u]$, and $\mathrm{Gal}(F_r)$ is a pro-p group if $p = \mathrm{char}(\bar{F}) \neq 0$.

Consider a finite extension $(L,v)/(F,v)$ of Henselian fields. Let $e = e(L/F) = \big(v(L^\times) : v(F^\times)\big)$ be the **ramification index**. There is a positive integer d such that $[L : F] = de[\bar{L} : \bar{F}]$. If $\mathrm{char}(\bar{F}) = p > 0$, then d is a power of p [Art, p. 62, Thm. 10]. If $\mathrm{char}(\bar{F}) = 0$, then $d = 1$. When $d = 1$, we say L/F is **defectless**. An arbitrary algebraic extension M/F is **defectless** if each finite subextension is defectless. This is the case when $\mathrm{char}(\bar{F}) \nmid [M : K]$. For example, F_r/F_u is defectless. In addition, by (2), $[L : F] = [\bar{L} : \bar{F}]$ for each finite subextension L/F of F_u/F. Hence, F_u/F is defectless. Consequently, F_r/F is defectless.

LEMMA 7.1. *Let (F,v) be a Henselian valued field. Use the above notation.*

(a) *There is a field F' with $F_uF' = F_r$ and $F_u \cap F' = F$.*

(b) *The short exact sequence $1 \to \mathrm{Gal}(F_r/F_u) \to \mathrm{Gal}(F_r/F) \to \mathrm{Gal}(F_u/F) \to 1$ splits.*

PROOF. Statement (b) is a Galois theoretic interpretation of (a). So, we prove (a).

Zorn's lemma gives a maximal extension F' of F in F_r with residue field $\bar{F}$. For each prime number $l \neq \mathrm{char}(\bar{F})$ the value group of F' is l-divisible. Otherwise, there is an $a \in F'$ with $v(a) \notin lv((F')^\times)$. Put $L = F'(\sqrt[l]{a})$. Then $[L : F'] = l$ and $l \le (v(L^\times) : v((F')^\times))$. Since $e(L/F')[\bar{L} : \bar{F'}] \le [L : F'] = l$, we have $\bar{L} = \bar{F'} = \bar{F}$. Recall: $\mathrm{Gal}(F_r)$ is a pro-p group if $\mathrm{char}(\bar{F}) = p > 0$ and trivial if $\mathrm{char}(\bar{F}) = 0$. Hence, $L \subseteq F_r$. This contradicts the maximality of F'.

By the discussion preceding Lemma 7.1, $F_u \cap F' = F$. Let $E = F_uF'$. Consider a prime number $l \neq \mathrm{char}(\bar{F})$. Since E/F' is an algebraic extension, $v(E^\times)$ is contained in the divisible hull of $v(F')$. Since $v((F')^\times)$ is l-divisible, so is $v(E^\times)$. Since $F_u \subseteq E \subseteq F_r \subseteq F_s$ and $\bar{F}_u = \bar{F}_s$, we have $\bar{E} = \bar{F}_r$. Hence, $e(E'/E) = [\bar{E'} : \bar{E}] = 1$ and therefore $[E' : E] = 1$ (because E'/E is defectless) for every finite extension E' of E in F_r. Consequently, $E = F_r$. □

LEMMA 7.2 (KUHLMANN-PANK-ROQUETTE [KPR, THM. 2.2]). *Let (F,v) be a Henselian field.*

(a) *There is a field F' with $F_r \cap F' = F$ and $F_rF' = F_s$.*

(b) *The short exact sequence $1 \to \mathrm{Gal}(F_r) \to \mathrm{Gal}(F) \to \mathrm{Gal}(F_r/F) \to 1$ splits.*

PROOF. Statement (a) is a Galois theoretic interpretation of (b). So, we prove (b). Let $p = \mathrm{char}(\bar{F})$. If $p = 0$, then $F_r = F_s$ and we may take $F' = F$. Suppose $p \neq 0$.

By (2), $\mathrm{Gal}(F_u/F) \cong \mathrm{Gal}(\bar{F})$. By Witt, the p-Sylow subgroups of $\mathrm{Gal}(\bar{F})$ are free [Rib, p. 256, Thm. 3.3]. Hence, so are the p-Sylow subgroups of $\mathrm{Gal}(F_u/F)$. Since $p \nmid [F_r : F_u]$, restriction maps each p-Sylow subgroup of $\mathrm{Gal}(F_r/F)$ isomorphically onto a p-Sylow subgroup of $\mathrm{Gal}(F_u/F)$. Hence, each p-Sylow subgroup of $\mathrm{Gal}(F_r/F)$ is free. Thus $\mathrm{cd}_p(\mathrm{Gal}(F_r/F)) = 1$ [Rib, p. 207, Cor. 2.2]. Since $\mathrm{Gal}(F_r)$ is a pro-p group, the short exact sequence in (b) splits [Rib, p. 211, Prop. 3.1(iii)'].□

PROPOSITION 7.3. *Let (F, v) be a valued field.*

(a) *Suppose (F, v) is Henselian. Then the epimorphism $\rho\colon \mathrm{Gal}(F) \to \mathrm{Gal}(\bar{F})$ induced by reduction at v splits.*

(b) *Each subgroup of* $\mathrm{Gal}(\bar{F})$ *is isomorphic to a subgroup of* $\mathrm{Gal}(F)$.

PROOF OF (a). The map ρ decomposes as $\mathrm{Gal}(F) \xrightarrow{\mathrm{res}} \mathrm{Gal}(F_r/F) \xrightarrow{\mathrm{res}} \mathrm{Gal}(F_u/F) \xrightarrow{\bar{\rho}} \mathrm{Gal}(\bar{F})$. The map $\bar{\rho}$ which is also induced by reduction is an isomorphism (by (2)). By Lemmas 7.1 and 7.2, each of the restriction maps splits. Hence ρ splits.

PROOF OF (b). Let (F', v) be the Henselization of (F, v). Then $\overline{F'} = \bar{F}$. By (a), each subgroup of $\mathrm{Gal}(\bar{F})$ is isomorphic to a subgroup of $\mathrm{Gal}(F')$, hence of $\mathrm{Gal}(F)$. □

PROPOSITION 7.4. *Let F/K be an extension of fields. Suppose v is a valuation of F which is trivial on K and $\bar{F} = K$. Then*

(a) $\mathrm{res}\colon \mathrm{Gal}(F) \to \mathrm{Gal}(K)$ *is an epimorphism which splits. If, in addition, (F, v) is Henselian and v is extended to F_s such that $\bar{a} = a$ for each $a \in K_s$, then* res *is the epimorphism induced by reduction at v.*

(b) *(F, v) has a separable algebraic Henselian extension (F', v) such that* $\mathrm{res}\colon \mathrm{Gal}(F') \to \mathrm{Gal}(K)$ *is an isomorphism and $\overline{F'}$ is a purely inseparable extension of K.*

(c) *Suppose K is perfect. Then (F, v) has an algebraic Henselian extension (F'', v) such that F'' is perfect, $\overline{F''} = K$, and* $\mathrm{res}\colon \mathrm{Gal}(F'') \to \mathrm{Gal}(K)$ *is an isomorphism.*

PROOF. Replace (F, v) by a Henselian closure, if necessary, to assume (F, v) is Henselian. Let $\rho\colon \mathrm{Gal}(F) \to \mathrm{Gal}(K)$ be the epimorphism induced by reduction at v. Then, for each $a \in K_s$ and each $\sigma \in \mathrm{Gal}(K)$ we have, $\sigma a = \overline{\sigma a} = \bar{\sigma}\bar{a} = \bar{\sigma} a = \rho(\sigma)a$. Thus, $\mathrm{res}\colon \mathrm{Gal}(F) \to \mathrm{Gal}(K)$ coincides with ρ.

Proposition 7.3(a) gives a section $\rho'\colon \mathrm{Gal}(K) \to \mathrm{Gal}(F)$ of ρ. Let F' be the fixed field of $\rho'(\mathrm{Gal}(K))$ in F_s. Then $\mathrm{Gal}(F') \to \mathrm{Gal}(K)$ is an isomorphism. Also, for all $u \in F'$ and $\sigma \in \mathrm{Gal}(F')$ we have $\bar{\sigma}\bar{u} = \overline{\sigma u} = \bar{u}$. Hence, $\overline{F'}$ is a purely inseparable extension of K. This concludes the proof of (a) and (b).

When K is perfect, $F'' = F'_{\mathrm{ins}}$ satisfies (c). □

The following Proposition gives more details to a result of Efrat [Efr. Prop. 4.7].

PROPOSITION 7.5. *Let K be a field, E_0 its prime field, and T a set of variables with* $\mathrm{card}(T) \geq \mathrm{trans.deg}(K/E_0)$. *Let F_0 be either E_0 or $\mathbb{Q}$. Then there is a field L, algebraic over $F_0(T)$, with $G(L) \cong G(K)$.*

PROOF. There is a unique place $\varphi_0: F_0 \to E_0 \cup \{\infty\}$. Choose a transcendence base $\bar{T}$ for K/E_0. By assumption, $\mathrm{card}(\bar{T}) \leq \mathrm{card}(T)$. Choose a surjective map $\varphi_1: T \to \bar{T}$. Then extend φ_0 and φ_1 to a place $\varphi: F_0(T) \to E_0(\bar{T}) \cup \{\infty\}$ and denote the corresponding valuation by v. Corollary 7.3(b) gives the desired field L. □

DEFINITION 7.6. *Rigid Henselian extensions.* Let K be a field and (L, v) a valued field. We say, (L, v) is a **rigid Henselian extension** of K if (L, v) is Henselian, $K \subseteq L$, v is trivial on K, $\bar{L}_v = K$, and $\mathrm{res}: \mathrm{Gal}(L) \to \mathrm{Gal}(K)$ is an isomorphism. In this case we also call the place $\varphi: L \to K \cup \{\infty\}$ associated with v **rigid**.

An arbitrary field extension L/K is a **rigid Henselian extension** if L admits a valuation v such that (L, v) is a rigid Henselian extension K. □

PROPOSITION 7.7. *Let F/K be a purely transcendental extension. Then:*
(a) *F has a valuation v which is trivial on K and $\bar{F} = K$.*
(b) *F has a separable algebraic extension F' such that* $\mathrm{res}: \mathrm{Gal}(F') \to \mathrm{Gal}(K)$ *is an isomorphism.*
(c) *If K is perfect, then F has a perfect algebraic extension F' which is a rigid Henselian extension of K.*

PROOF OF (a). The assertion is evident when $F = K(t)$ and t is transcendental. The general case follows from the special case by transfinite induction and using composition of valuations.

PROOF OF (b). Apply Proposition 7.4(b).

PROOF OF (c). Apply Proposition 7.4(c). □

8. The Space of Valuations of a Field

Let K be a field. Denote the collection of all valuations of K by $\mathrm{Val}(K)$. We include in $\mathrm{Val}(K)$ also the trivial valuation v_0 defined by $v_0(a) = 0$ for each $a \in K^\times$ and $v_0(0) = \infty$. We do not distinguish between equivalent valuations. Thus, we identify valuations with the same valuation rings. Given $a \in K$, we write

$$\mathrm{Val}_a(K) = \{v \in \mathrm{Val}(K) \mid v(a) > 0\}, \qquad \mathrm{Val}'_a(K) = \{v \in \mathrm{Val}(K) \mid v(a) \geq 0\}.$$

Intersections of finitely many sets of these form build a basis for a topology on $\mathrm{Val}(K)$, the so called **patch topology** (see more about the patch topology in [Hoe, Sec. 2]).

The following identities make the use of open subsets of $\mathrm{Val}(K)$ easier:

$$\begin{aligned} &\mathrm{Val}_{a/b}(K) = \{v \in K \mid v(a) > v(b)\}, \qquad \mathrm{Val}'_{a/b}(K) = \{v \in K \mid v(a) \geq v(b)\}, \\ &\mathrm{Val}'_a(K) = \mathrm{Val}(K) \smallsetminus \mathrm{Val}_{a^{-1}}(K), \quad \mathrm{Val}_0(K) = \mathrm{Val}(K), \quad \mathrm{Val}_1(K) = \emptyset. \end{aligned} \tag{1}$$

EXAMPLE 8.1. $\mathrm{Val}(\mathbb{Q})$. It consists of v_p, with p ranging over all prime numbers, and v_0. For each p, $\{v \in \mathrm{Val}(\mathbb{Q}) \mid v(p) > 0\} = \{v_p\}$. Thus v_p is a discrete point of $\mathrm{Val}(\mathbb{Q})$. On the other hand, $v_0(a) = 0$ for each $a \in \mathbb{Q}^\times$. Hence, if $\mathbf{B} = \bigcap_{i=1}^m \mathrm{Val}_{a_i}(\mathbb{Q}) \cap \bigcap_{j=1}^n \mathrm{Val}'_{b_j}(\mathbb{Q})$ contains v_0, then we may assume $m = 0$. Hence, $\mathbf{B}$ contains all v_p with p relatively prime to all denominators of b_j. This implies, every open neighborhood of v_0 consists of almost all elements of $\mathrm{Val}(\mathbb{Q})$. Hence, $\mathrm{Val}(\mathbb{Q})$ consists of a discrete sequence converging to v_0. In particular, $\mathrm{Val}(\mathbb{Q})$ is compact. □

The following result generalizes the last conclusion of Example 8.1.

PROPOSITION 8.2. *$\mathrm{Val}(K)$ is profinite.*

PROOF. The space $\mathrm{Sign}(K) = \prod_{a\in K^\times}\{-1,0\}$ with the product topology is a profinite space. For each $v \in \mathrm{Val}(K)$ and $a \in K^\times$ let $\mathrm{sign}(v(a))$ be -1 if $v(a) < 0$ and 0 if $v(a) \geq 0$. Define a map $\sigma\colon \mathrm{Val}(K) \to \mathrm{Sign}(K)$ by $\sigma(v)(a) = \mathrm{sign}(v(a))$. It suffices to prove that σ is a homeomorphism onto a closed subset of $\mathrm{Sign}(K)$.

Indeed, let $v, v' \in \mathrm{Val}(K)$ with $\sigma(v) = \sigma(v')$. Then $v(a) \geq 0$ if and only if $v'(a) \geq 0$. Hence, $v = v'$. Therefore, σ is injective.

A basic open subset of $\sigma(\mathrm{Val}(K))$ has the form

$$\{\sigma(v) \mid v \in \mathrm{Val}(K),\ \mathrm{sign}(v(a_i)) = -1,\ i = 1, \dots, m,\ \mathrm{sign}(v(b_j)) = 0,\ j = 1, \dots, n\}$$

with $a_1, \dots, a_m, b_1, \dots, b_n \in K^\times$ and $m, n \geq 0$. It is the image of the basic open subset $\bigcap_{i=1}^m \mathrm{Val}_{a_i^{-1}}(K) \cap \bigcap_{j=1}^n \mathrm{Val}'_{b_j}(K)$. Therefore, σ is a homeomorphism.

Next consider an element $f \in \mathrm{Sign}(K)$ which belongs to the closure of $\mathrm{Im}(\sigma)$. We construct $w \in \mathrm{Val}(K)$ with $\sigma(w) = f$. This will conclude the proof of the proposition. Put $O = \{a \in K^\times \mid f(a) = 0\} \cup \{0\}$.

CLAIM. *O is a valuation ring.* Indeed, assume $a, b \in O$ but $a + b \notin O$. Then $a, b \neq 0$ and $\{g \in \mathrm{Sign}(K) \mid g(a) = 0,\ g(b) = 0,\ g(a+b) = -1\}$ is an open neighborhood of f. Hence, there is a $v \in \mathrm{Val}(K)$ with $\mathrm{sign}(v(a)) = 0$, $\mathrm{sign}(v(b)) = 0$, and $\mathrm{sign}(v(a+b)) = -1$. Thus, $v(a) \geq 0$, $v(b) \geq 0$, and $v(a+b) < 0$. This contradiction proves that O is closed under addition.

Similarly, O is closed under multiplication and contains $0, -1$. Hence, O is a subring of K.

Now let $a \in K^\times$. If $f(a) = f(a^{-1}) = -1$, there is a $v \in \mathrm{Val}(K)$ with $v(a) < 0$ and $v(a^{-1}) < 0$, a contradiction. Hence, $a \in O$ or $a^{-1} \in O$. Therefore, O is a valuation ring.

Denote the valuation associated with O by w. Then $\mathrm{sign}(w(a)) = f(a)$ for each $a \in K$. Therefore, $f = \sigma(w)$. □

For a valued field (K, v) and a polynomial $f(X) = \sum_{i=0}^n a_i X^i$ with $a_i \in K$ we write $v(f) = \min(v(a_0), \dots, v(a_n))$. Also, for $\mathbf{x} = (x_1, \dots, x_n) \in K^n$ we write $v(\mathbf{x}) = \min(v(x_1), \dots, v(x_n))$.

LEMMA 8.3. *Let (K, v) be a valued field, (K_v, v_h) be a Henselian closure of (K, v), and L a finite separable extension of K. Then the following conditions are equivalent:*

(a) *There is a K-embedding of L into K_v.*

(b) *L/K has a primitive element x such that*

$$\mathrm{irr}(x, K) = X^n + X^{n-1} + a_{n-2}X^{n-2} + \cdots + a_0$$

with $v(a_i) > 0$, $i = 0, \dots, n-2$.

PROOF OF "(a) $\Longrightarrow$ (b)". The embedding of L into K_v induces a valuation v_L of L which extends v. Let $\hat{L}$ be the Galois closure of L/K. Choose an extension $\hat{v}$ of v_L to $\hat{L}$. Then, L is contained in the decomposition field L' of $\hat{v}$ over K. Hence, $\mathrm{Gal}(\hat{L}/L') \leq \mathrm{Gal}(\hat{L}/L)$.

Every extension of v to $\hat{L}$ has the form $\hat{v} \circ \sigma$ with $\sigma \in \mathrm{Gal}(\hat{L}/K)$. We have, $\mathrm{res}_L(\hat{v} \circ \sigma) = \mathrm{res}_L \hat{v}$ if and only if there is a $\tau \in \mathrm{Gal}(\hat{L}/L)$ with $\hat{v} \circ \sigma = \hat{v} \circ \tau$, that is,

$\sigma\tau^{-1}$ lies in the decomposition group $\mathrm{Gal}(\hat{L}/L')$ of $\hat{v}$. Conclude: $\mathrm{res}_L(\hat{v}\circ\sigma) = \mathrm{res}_L\hat{v}$ if and only if $\sigma \in \mathrm{Gal}(\hat{L}/L)$.

Now use the Chinese remainder theorem [Jar, Lemma 6.7(c)] to find $y \in L$ with $\hat{v}(y) = 0$ and $\hat{v}(\sigma y) > 0$ for each $\sigma \in \mathrm{Gal}(\hat{L}/K) \smallsetminus \mathrm{Gal}(\hat{L}/L)$. Next choose a primitive element z for L/K. Multiply z by a suitable element of K to assume

$$\hat{v}(\sigma z) > \max(0, \hat{v}(y' - y)) \tag{2}$$

for each $\sigma \in \mathrm{Gal}(\hat{L}/K)$ and every conjugate y' of y over K with $y' \neq y$. Put $x = y + z$. Then

$$\hat{v}(x) = 0 \text{ and } \hat{v}(\sigma x) > 0 \text{ for each } \sigma \in \mathrm{Gal}(\hat{L}/K) \smallsetminus \mathrm{Gal}(\hat{L}/L). \tag{3}$$

We prove $L = K(x)$.

To this end consider $\tau \in \mathrm{Gal}(\hat{L}/K(x))$. Then $\tau(y) - y = z - \tau(z)$. Therefore,

$$\hat{v}(\tau(y) - y) \geq \min(\hat{v}(z), \hat{v}(\tau(z))) \geq \min_{\sigma\in\mathrm{Gal}(\hat{L}/K)} \hat{v}(\sigma z).$$

By (2), $\tau(y) = y$. Hence, $\tau(z) = z$. Therefore, $L = K(z) \subseteq K(x) \subseteq L$. It follows that $L = K(x)$, as contended.

Let $x_1, \dots, x_n$ be the conjugates of x in L with $x_1 = x$. For each $j \geq 2$ there is a $\sigma \in \mathrm{Gal}(\hat{L}/K) \smallsetminus \mathrm{Gal}(\hat{L}/L)$ with $\sigma x = x_j$. Hence, by (3),

$$\hat{v}(x_1) = 0 \text{ and } \hat{v}(x_j) > 0 \text{ if } j > 2. \tag{4}$$

Let $f(X) = X^n + b_{n-1}X^{n-1} + b_{n-2}X^{n-2} + \cdots + b_0 = \mathrm{irr}(x, K)$. By (4), $\hat{v}(b_{n-1}) = \hat{v}(x_1 + x_2 + \cdots + x_n) = 0$, $\hat{v}(b_{n-2}) = \hat{v}(\sum_{j\neq k} x_j x_k) > 0$, $\cdots$, $\hat{v}(b_0) = \hat{v}(x_1 \cdots x_n) > 0$. Obviously, $\frac{x}{b_{n-1}}$ is a primitive element for L/K. Its irreducible polynomial over K is

$$X^n + X^{n-1} + \frac{b_{n-2}}{b_{n-1}^2} X^{n-2} + \cdots + \frac{b_0}{b_{n-1}^n}.$$

This polynomial has the required form.

PROOF OF "(b) $\Longrightarrow$ (a)". In the notation of (b) let $f = \mathrm{irr}(x, K)$. Then $v(f(-1)) > 0$ and $v(f'(-1)) = v((-1)^{n-1}) = 0$. Hence, by Hensel's Lemma, f has a root $x' \in K_v$. The map $x \mapsto x'$ extends to a K-embedding of L into K_v. □

LEMMA 8.4. (OPEN MAP THEOREM) *Let L be a field extension of K. Then the map $\mathrm{res}_{L/K}\colon \mathrm{Val}(L) \to \mathrm{Val}(K)$ is continuous. If L/K is separable algebraic, then the map is also open.*

PROOF. By definition, $\mathrm{res}_{L/K}^{-1}(\mathrm{Val}_a(K)) = \mathrm{Val}_a(L)$ and $\mathrm{res}_{L/K}^{-1}(\mathrm{Val}'_a(K)) = \mathrm{Val}'_a(L)$ for each $a \in K^\times$. Hence, restriction of valuations of L to K is a continuous map.

Now suppose L/K is Galois. Put $G = \mathrm{Gal}(L/K)$. Then G acts on $\mathrm{Val}(L)$ continuously and $\mathrm{res}_{L/K}$ induces a continuous bijective map $\rho\colon \mathrm{Val}(L)/G \to \mathrm{Val}(K)$. Since both spaces are profinite, ρ is a homeomorphism. By definition, the quotient map $\pi\colon \mathrm{Val}(L) \to \mathrm{Val}(L)/G$ is open. Thus, $\mathrm{res}_{L/K} = \rho \circ \pi$ is also open.

Finally suppose L/K is separable algebraic. Let $\hat{L}$ be the Galois closure of L/K. Then $\mathrm{res}_{\hat{L}/L}$ is continuous and $\mathrm{res}_{\hat{L}/K}$ is open. Let U be an open subset of $\mathrm{Val}(L)$. Then $\mathrm{res}_{L/K}(U) = \mathrm{res}_{\hat{L}/K}(\mathrm{res}_{\hat{L}/L}^{-1}(U))$ is open, as desired. □

9. Locally Uniform v-adic Topologies

Every valuation v of a field K gives rise to a topology on K which naturally extends to a topology on $V(K)$ (called the **v-topology**) for every variety V defined over K. Polynomials $f \in K[X]$ and in general morphisms between varieties over K are continuous in the v-topology. The proof of continuity uses only finitely many conditions of the form $v(a) > 0$ and $v(a') \geq 0$. Therefore, it holds for all valuations v' of K satisfying the same conditions. In other words, polynomials are "locally uniform continuous". This observation holds even if we consider the polynomials as functions of valued fields extending (K, v).

The aim of this section is the make this heuristic argument precise. It will be used in Proposition 12.4 to prove that every field-valuation structure satisfying the block approximation condition is unirationally closed.

We start by choosing a **large universal extension** of K. This is an algebraically closed field extension Ω of K with $\mathrm{trans.deg}(\Omega/K) > \mathrm{card}(K)$. Denote the set of all field extensions L of K with $L \subseteq \Omega$ and $\mathrm{trans.deg}(L/K) \leq \mathrm{card}(K)$ by $\mathrm{Exten}(K)$. For each $v \in \mathrm{Val}(K)$ denote the set of all valued fields (L, w) extending (K, v) with $L \in \mathrm{Exten}(K)$ by $\mathrm{Exten}(K, v)$. For each subset $\mathbf{B}$ of $\mathrm{Val}(K)$ let $\mathrm{Exten}(K, \mathbf{B}) = \bigcup_{v\in\mathbf{B}} \mathrm{Exten}(K, v)$. In addition, let $\mathrm{Hensel}(K, \mathbf{B})$ be the set of all Henselian fields (L, w) in $\mathrm{Exten}(K, \mathbf{B})$.

The reason for working inside Ω is to avoid using classes, especially to avoid operations with classes which may led to set theoretic paradoxes.

Denote the collection of all subsets of a set A by $\mathrm{Subset}(A)$. Consider a reduced scheme of finite type V over K and a subset $\mathbf{B}$ of $\mathrm{Val}(K)$. Let

$$\mathrm{Set}(K, V, \mathbf{B}) = \prod_{(L,v)\in\mathrm{Exten}(K,\mathbf{B})} \mathrm{Subset}(V(L)).$$

Thus, each element of $\mathrm{Set}(K, V, \mathbf{B})$ is a set valued function $\mathcal{V}$ from $\mathrm{Exten}(K, \mathbf{B})$ satisfying $\mathcal{V}(L, v) \subseteq V(L)$ for each $(L, v) \in \mathrm{Exten}(K, \mathbf{B})$. Regard V itself as an element of $\mathrm{Set}(K, V, \mathbf{B})$.

Let $\mathcal{V}, \mathcal{V}' \in \mathrm{Set}(K, V, \mathbf{B})$. We write $\mathcal{V} \subseteq \mathcal{V}'$ if $\mathcal{V}(L, v) \subseteq \mathcal{V}'(L, v)$ for all $(L, v) \in \mathrm{Exten}(K, \mathbf{B})$,

The **restriction** of $\mathcal{V}$ to a subset $\mathbf{B}_0$ of $\mathbf{B}$ is the function $\mathcal{V}|_{\mathbf{B}_0} \in \mathrm{Set}(K, V, \mathbf{B}_0)$ defined by $\mathcal{V}|_{\mathbf{B}_0}(L, v) = \mathcal{V}(L, v)$ for each $L \in \mathrm{Exten}(K, \mathbf{B}_0)$.

Define **unions** and **intersections** in $\mathrm{Set}(K, V, \mathbf{B})$ via unions and intersections of sets:

$$\Big(\bigcup_{i\in I} \mathcal{V}_i\Big)(L, v) = \bigcup_{i\in I} \mathcal{V}_i(L, v) \qquad \Big(\bigcap_{i\in I} \mathcal{U}_i\Big)(L, v) = \bigcap_{i\in I} \mathcal{U}_i(L, v)$$

These operations satisfy the usual de-Morgan laws. Similarly define the direct product of $\mathcal{U} \in \mathrm{Set}(K, U, \mathbf{B})$ with $\mathcal{V} \in \mathrm{Set}(K, V, \mathbf{B})$ by the rule $(\mathcal{U} \times \mathcal{V})(L, v) = \mathcal{U}(L, v) \times \mathcal{V}(L, v)$.

Let $\mathbf{a} \in K^n$, $\mathbf{c} \in (K^\times)^m$, and $f_1, \ldots, f_m \in K[X_1, \ldots, X_n]$. Define an element $\mathcal{O}_{\mathbf{a},\mathbf{c},\mathbf{f},\mathbf{B}}$ in $\mathrm{Set}(K, \mathbb{A}^n, \mathbf{B})$ in the following way: For all $(L, v) \in \mathrm{Exten}(K, \mathbf{B})$

$$\mathcal{O}_{\mathbf{a},\mathbf{c},\mathbf{f},\mathbf{B}}(L, v) = \{\mathbf{x} \in L^n \mid v(f_i(\mathbf{x}) - f_i(\mathbf{a})) > v(c_i),\ i = 1, \ldots, m\}.$$

Note that $\mathcal{O}_{\mathbf{a},\mathbf{c},\mathbf{f},\mathbf{B}}(L, v)$ is a v-open neighborhood of $\mathbf{a}$ in L^n. If we embed K^n diagonally in $\prod_{(L,v)\in\mathrm{Exten}(K,\mathbf{B})} L^n$, then $\mathbf{a}$ belongs to $\prod_{(L,v)\in\mathrm{Exten}(K,\mathbf{B})} \mathcal{O}_{\mathbf{a},\mathbf{c},\mathbf{f},\mathbf{B}}(L, v)$. Hence, we call $\mathbf{O}_{\mathbf{a},\mathbf{c},\mathbf{f},\mathbf{B}}$ a **basic open neighborhood** of $\mathbf{a}$ in $\mathrm{Set}(K, \mathbb{A}^n_K, \mathbf{B})$. The intersection of finitely many basic open neighborhoods of $\mathbf{a}$ is again a basic open neighborhood of $\mathbf{a}$ in $\mathrm{Set}(K, \mathbb{A}_K^n, \mathbf{B})$. Define an **open neighborhood** of $\mathbf{a}$ in $\mathrm{Set}(K, \mathbb{A}_K^n, \mathbf{B})$ to be a union of basic open neighborhoods of $\mathbf{a}$ in $\mathrm{Set}(K, \mathbb{A}_K^n, \mathbf{B})$.

An example of an open neighborhood of $\mathbf{a}$ in $\mathrm{Set}(K, \mathbb{A}_K^n, \mathbf{B})$ is an **open ball**:

$$\mathcal{B}_{\mathbf{a},c,\mathbf{B}}(L, v) = \{\mathbf{x} \in L^n \mid v(\mathbf{x} - \mathbf{a}) > v(c)\}.$$

Let V be a Zariski closed subset of $\mathbb{A}_K^n$, $\mathbf{a} \in V(K)$, and $\mathcal{V}$ an open neighborhood of $\mathbf{a}$ in $\mathrm{Set}(K, \mathbb{A}_K^n, \mathbf{B})$. Refer to $V \cap \mathcal{V}$ as an **open neighborhood** of $\mathbf{a}$ in $\mathrm{Set}(K, V, \mathbf{B})$.

REMARK 9.1. Let $v \in \mathrm{Val}(K)$, $\mathbf{a} \in K^n$, and $c, d \in K$. Suppose $v(c) \leq v(d)$. Then $\mathbf{B} = \{w \in \mathrm{Val}(K) \mid w(c) \leq w(d)\}$ is an open neighborhood of v in $\mathrm{Val}(K)$. Moreover, $\mathcal{B}_{\mathbf{a},d,\mathbf{B}}(L, w) \subseteq \mathcal{B}_{\mathbf{a},c,\mathbf{B}}(L, w)$ for all $(L, w) \in \mathrm{Exten}(K, \mathbf{B})$. □

DEFINITION 9.2. *Uniform local topology on schemes.* Let V be a Zariski closed subset of $\mathbb{A}_K^m$, W a Zariski closed subset in $\mathbb{A}_K^n$, and $\varphi: V \to W$ a K-morphism. Then there are polynomials $f_1, \dots, f_n \in K[X_1, \dots, X_m]$ with $\varphi(\mathbf{x}) = (f_1(\mathbf{x}), \dots, f_n(\mathbf{x}))$ for all $L \in \mathrm{Exten}(K)$ and $\mathbf{x} \in V(L)$.

Let $\mathbf{B}$ be a subset of $\mathrm{Val}(K)$. For each $\mathcal{V} \in \mathrm{Set}(K, V, \mathbf{B})$ define $\varphi(\mathcal{V})$ to be the element of $\mathrm{Set}(K, W, \mathbf{B})$ given by $\varphi(\mathcal{V})(L, v) = \varphi(\mathcal{V}(L, v))$. Similarly, for each $\mathcal{W} \in \mathrm{Set}(K, W, \mathbf{B})$ define $\varphi^{-1}(\mathcal{W})$ to be the element of $\mathrm{Set}(K, V, \mathbf{B})$ defined by $\varphi^{-1}(\mathcal{W})(L, v) = \varphi^{-1}(\mathcal{W}(L, v))$.

As an example, let $\mathbf{a} \in V(K)$, $\mathbf{b} = \varphi(\mathbf{a})$, and $g_1, \dots, g_k \in K[Y_1, \dots, Y_n]$. Then $\mathbf{g} \circ \varphi = (h_1, \dots, h_k)$ with $h_i(\mathbf{X}) = g_i(f_1(\mathbf{X}), \dots, f_n(\mathbf{X}))$ and

$$\varphi^{-1}(W \cap \mathcal{O}_{\mathbf{b},\mathbf{c},\mathbf{g},\mathbf{B}}) = V \cap \mathcal{O}_{\mathbf{a},\mathbf{c},\mathbf{g}\circ\varphi,\mathbf{B}}.$$

Hence, the inverse image under φ of any open neighborhood of $\mathbf{b}$ in $\mathrm{Set}(K, W, \mathbf{B})$ is an open neighborhood of $\mathbf{a}$ in $\mathrm{Set}(K, V, \mathbf{B})$. In particular, if φ is an isomorphism, $\mathcal{V}$ is an open neighborhood of $\mathbf{a}$ in $\mathrm{Set}(K, V, \mathbf{B})$, and $\mathcal{W} = \varphi(\mathcal{V})$, then $\mathcal{W}$ is an open neighborhood of $\mathbf{b}$ in $\mathrm{Set}(K, W, \mathbf{B})$ and $\varphi^{-1}(\mathcal{W}) = \mathcal{V}$.

Now let V be a reduced scheme of finite type over K and $\mathbf{a} \in V(K)$. Choose a Zariski K-open affine neighborhood V_0 of $\mathbf{a}$ in V. Each open neighborhood of $\mathbf{a}$ in $\mathrm{Set}(K, V_0, \mathbf{B})$ is an **open neighborhood** of $\mathbf{a}$ in $\mathrm{Set}(K, V, \mathbf{B})$. The observation of the preceding paragraph shows this definition is independent of V_0. □

LEMMA 9.3. (LOCAL UNIFORM CONTINUITY OF POLYNOMIALS) *Let (K, v) be a valued field, $g \in K[X_1, \dots, X_n]$, $\mathbf{a}, \mathbf{x} \in K^n$, and $e \in K^\times$. Suppose $v(g) \geq 0$, $v(\mathbf{a}, \mathbf{x}) \geq 0$, and $v(\mathbf{x} - \mathbf{a}) > v(e)$. Then $v(g(\mathbf{x}) - g(\mathbf{a}))) > v(e)$.*

PROOF. We prove the Lemma by induction on n.

First suppose $n = 1$. Write $g(X) = \sum_{i=0}^r c_i X^i$ with $c_i \in K$ satisfying $v(c_i) \geq 0$, $i = 0, \dots, r$. Then

$$\begin{aligned} v(g(x) - g(a)) &= v\Big(\sum_{i=0}^r c_i(x^i - a^i)\Big) \\ &\geq \min_{1 \leq i \leq r} (v(c_i) + v(x - a) + v(x^{i-1} + x^{i-2}a + \cdots + a^{i-1})) \\ &> v(e). \end{aligned}$$

Now assume $n > 2$ and the statement holds up to $n-1$. Then

$$\begin{aligned} v(g(\mathbf{x}) - g(\mathbf{a})) \geq \min \big(& v(g(x_1, \dots, x_{n-1}, x_n) - g(x_1, \dots, x_{n-1}, a_n)), \\ & v(g(x_1, \dots, x_{n-1}, a_n) - g(a_1, \dots, a_{n-1}, a_n))\big) > v(e). \end{aligned}$$

This concludes the induction. □

As a consequence we show that open balls are locally basic open neighborhoods of K-rational points on varieties over K.

LEMMA 9.4. *Let K be a field, V a Zariski closed subset of $\mathbb{A}^n_K$, $\mathbf{a} \in V(K)$, $\mathbf{B}$ a closed subset of $\mathrm{Val}(K)$, and $\mathcal{V}$ an open neighborhood of $\mathbf{a}$ in $\mathrm{Set}(K, V, \mathbf{B})$. Then there is a partition $\mathbf{B} = \bigcup_{i=1}^m \mathbf{B}_i$ with $\mathbf{B}_i$ closed and for each i there is an open ball $\mathcal{B}_{\mathbf{a},c_i,\mathbf{B}_i}$ in $\mathrm{Set}(K, \mathbb{A}^n_K, \mathbf{B}_i)$ such that $V(L) \cap \mathcal{B}_{\mathbf{a},c_i,\mathbf{B}_i}(L, v) \subseteq \mathcal{V}(L, v)$ for each $(L, v) \in \mathrm{Exten}(K, \mathbf{B}_i)$.*

PROOF. Assume without loss $V = \mathbb{A}^n_K$. Choose $c'_1, \dots, c'_l \in K^\times$ and $f_1, \dots, f_l \in K[X_1, \dots, X_n]$ such that $\mathcal{O}_{\mathbf{a},\mathbf{c}',\mathbf{f},\mathbf{B}}$ is an open neighborhood of $\mathbf{a}$ in $\mathcal{V}$. For each $v \in \mathbf{B}$ choose $e_v \in K^\times$ with $v(e_v \mathbf{a}) \geq 0$. Put $g_{v,k}(\mathbf{X}) = f_k(\frac{1}{e_v}\mathbf{X})$, $k = 1, \dots, l$. Next choose $d_v \in K^\times$ with $v(d_v g_{v,k}) \geq 0$ for $k = 1, \dots, l$. Finally choose $c_v \in K^\times$ with $v(c_v e_v) \geq 0$ and $v\Big(\frac{c_v e_v}{d_v c'_k}\Big) \geq 0$ for $k = 1, \dots, l$. Then

$$\begin{aligned} \mathbf{B}'_v = \{v' \in \mathbf{B} \mid v'(e_v \mathbf{a}) \geq 0,\ v'(d_v g_{v,k}) \geq 0,\ v'(c_v e_v) \geq 0,\ v'\Big(\frac{c_v e_v}{d_v c'_k}\Big) \geq 0, \\ k = 1, \dots, l\} \end{aligned}$$

is an open neighborhood of v in $\mathbf{B}$.

By Lemma 8.2, $\mathbf{B}$ is profinite. Hence, $\mathbf{B}'_v$ has a subset $\mathbf{B}_v$ which is open-closed in $\mathbf{B}$ and contains v. Compactness of $\mathbf{B}$ gives $v_1, \dots, v_m \in \mathbf{B}$ with $\mathbf{B} = \bigcup_{i=1}^m \mathbf{B}_{v_i}$. Let $\mathbf{B}_1 = \mathbf{B}_{v_1}$ and $\mathbf{B}_i = \mathbf{B}_{v_i} \smallsetminus (\mathbf{B}_{v_1} \cup \dots \cup \mathbf{B}_{v_{i-1}})$, $i = 2, \dots, m$. Then $\mathbf{B}_i$ is closed in $\mathrm{Val}(K)$, $\mathbf{B}_i \subseteq \mathbf{B}'_{v_i}$, $i = 1, \dots, m$, and $\mathbf{B} = \bigcup_{i=1}^m \mathbf{B}_i$.

Now consider an i between 1 and m. Put $c_i = c_{v_i}$, $d_i = d_{v_i}$, $e_i = e_{v_i}$, and $g_{ik} = g_{v_i,k}$ for $k = 1, \dots, l$. It suffices to prove that

$$\mathcal{B}_{\mathbf{a},c_i,\mathbf{B}_i}(L, w) \subseteq \mathcal{O}_{\mathbf{a},\mathbf{c}',\mathbf{f},\mathbf{B}_i}(L, w)$$

for each $(L, w) \in \mathrm{Exten}(K, \mathbf{B}_i)$.

Indeed, our choices imply

(5) $\quad w(e_i \mathbf{a}) \geq 0,\ w(d_i g_{ik}) \geq 0,\ w(c_i e_i) \geq 0,\ w(c_i e_i) \geq w(d_i c'_k),\ k = 1, \dots, l.$

Let $\mathbf{x} \in \mathcal{B}_{\mathbf{a},\mathbf{c},\mathbf{B}_i}(L, w)$. Then $w(\mathbf{x} - \mathbf{a}) > w(c_i)$. Hence, by (5), $w(e_i \mathbf{x} - e_i \mathbf{a}) > w(c_i e_i) \geq 0$. Therefore, by (5), $w(e_i \mathbf{x}) \geq 0$. It follows from (5) and Lemma 9.3 that

$$w(d_i g_{ik}(e_i \mathbf{x}) - d_i g_{ik}(e_i \mathbf{a})) > w(c_i e_i) \geq w(d_i c'_k),\ k = 1, \dots, l.$$

Thus, $w(f_k(\mathbf{x}) - f_k(\mathbf{a})) > w(c'_k)$, $k = 1, \dots, l$. This means $\mathbf{x} \in \mathcal{B}_{\mathbf{a},\mathbf{c}',\mathbf{f},\mathbf{B}_i}(L, w)$, as claimed. □

10. Locally Uniform Hensel's Lemma

Let (K, v) be a valued field, $\varphi: V \to W$ a morphism of absolutely irreducible varieties over K, $\mathbf{a} \in V_{\mathrm{simp}}(K)$, $\mathbf{b} \in W_{\mathrm{simp}}(K)$, and $\varphi(\mathbf{a}) = \mathbf{b}$. Suppose φ is étale at $\mathbf{a}$. Let (L, v) be a Henselian extension of (K, v). Then $\mathbf{a}$ has a v-open neighborhood $\mathcal{V}$ in $V(L)$ and $\mathbf{b}$ has a v-open neighborhood $\mathcal{W}$ in $W(L)$ such that $\varphi: \mathcal{V}(L) \to \mathcal{W}(L)$ is a v-homeomorphism [GPR, Thm. 9.4]. The proof of this result relies on a higher dimensional Hensel's Lemma.

We strengthen this result by making $\mathcal{V}$ and $\mathcal{W}$ uniform on an open neighborhood of v in $\mathrm{Val}(K)$. The proof reduces the general case to the case where V is a hypersurface in $\mathbb{A}_K^{r+1}$, $W = \mathbb{A}_K^r$, and φ is the projection on the first r coordinates. Then we use a sharp form of Hensel's lemma.

LEMMA 10.1. *Let (L, w) be a Henselian field and $f \in L[T_1, \dots, T_r, X]$ monic in X. Put $f' = \frac{\partial f}{\partial X}$. Assume $w(f) \geq 0$ (hence $w(f') \geq 0$). Let $\mathbf{b}_0, \mathbf{b} \in L^r$, $c_0 \in L$, and $\epsilon \geq \delta \geq 0$ be in $w(L^\times)$. Suppose*

$$w(\mathbf{b}_0, c_0) \geq 0, \tag{1a}$$

$$w\big(f'(\mathbf{b}_0, c_0)\big) = \delta, \tag{1b}$$

$$w\big(f(\mathbf{b}_0, c_0)\big) > \delta + \epsilon, \text{ and} \tag{1c}$$

$$w(\mathbf{b} - \mathbf{b}_0) > \delta + \epsilon. \tag{1d}$$

Then $w(\mathbf{b}) \geq 0$ and there is a unique $c \in L$ with $f(\mathbf{b}, c) = 0$ and $w(c - c_0) > \epsilon$. In particular, $w(c) \geq 0$ and $w\big(f'(\mathbf{b}, c)\big) = \delta$.

PROOF. By (1a) and (1d), $w(\mathbf{b}) \geq 0$. By (1d) and Lemma 9.3

$$w\big(f'(\mathbf{b}, c_0) - f'(\mathbf{b}_0, c_0)\big) > \delta + \epsilon \geq \delta, \tag{2a}$$

$$w\big(f(\mathbf{b}, c_0) - f(\mathbf{b}_0, c_0)\big) > \delta + \epsilon, \tag{2b}$$

Hence, by (1a) and (1c)

$$w\big(f'(\mathbf{b}, c_0)\big) = \delta, \qquad w\big(f(\mathbf{b}, c_0)\big) > \delta + \epsilon = 2\delta + (\epsilon - \delta). \tag{3}$$

A sharp form of Hensel's lemma [Jar, Prop. 11.1(e)] gives a unique $c \in L$ such that $f(\mathbf{b}, c) = 0$ and $w(c - c_0) > \delta + (\epsilon - \delta) = \epsilon \geq \delta$. By (1a), $w(c) \geq 0$. By (1d) and Lemma 9.3, $w(f'(\mathbf{b}, c) - f'(\mathbf{b}_0, c_0)) > \delta$. Hence, by (1b), $w\big(f'(\mathbf{b}, c)\big) = \delta$. $\square$

For each $f \in K[X_1, \dots, X_n]$ let $V(f)$ be the hypersurface in $\mathbb{A}_K^n$ defined by $f = 0$.

LEMMA 10.2. *Let $f \in K[T_1, \dots, T_r, X]$, $v \in \mathrm{Val}(K)$, and $(\mathbf{b}_0, c_0) \in K^{r+1}$. Put $V = V(f)$ and $f' = \frac{\partial f}{\partial X}$. Suppose f is monic in X,*

$$v(f) \geq 0, \ v(\mathbf{b}_0, c_0) \geq 0, \quad \text{and} \quad v(f(\mathbf{b}_0, c_0)) > 2v(f'(\mathbf{b}_0, c_0)). \tag{4}$$

Then v has an open neighborhood $\mathbf{B}$ in $\mathrm{Val}(K)$, $\mathbf{b}_0$ has an open neighborhood $\mathcal{B}$ in $\mathrm{Set}(K, \mathbb{A}_K^r, \mathbf{B})$, and c_0 has an open neighborhood $\mathcal{C}$ in $\mathrm{Set}(K, \mathbb{A}_K^1, \mathbf{B})$ satisfying this: For each $(L, w) \in \mathrm{Hensel}(K, \mathbf{B})$ the projection

$$\mathrm{pr}: (\mathcal{B}(L, w) \times \mathcal{C}(L, w)) \cap V(L) \to \mathcal{B}(L, w) \tag{5}$$

is a w-homeomorphism.

PROOF. The sharp inequality in (4) implies $f'(\mathbf{b}_0, c_0) \neq 0$. Hence,

$$\mathbf{B} = \{w \in \mathrm{Val}(K) \mid w(f) \geq 0,\ w(\mathbf{b}_0, c_0) \geq 0, \text{ and } w(f(\mathbf{b}_0, c_0)) > 2w(f'(\mathbf{b}_0, c_0))\}$$

is an open neighborhood of v in $\mathrm{Val}(K)$.

Consider $(L, w) \in \mathrm{Hensel}(K, \mathbf{B})$. Let $\delta = w(f'(\mathbf{b}_0, c_0))$. Then

$$w(f) \geq 0, \quad w(\mathbf{b}_0, c_0) \geq 0, \quad \text{and} \quad w(f(\mathbf{b}_0, c_0)) > 2\delta.$$

Let

$$\mathcal{B}(L, w) = \{\mathbf{b} \in L^r \mid w(\mathbf{b} - \mathbf{b}_0) > 2\delta\} \quad \text{and} \quad \mathcal{C}(L, w) = \{c \in L \mid w(c - c_0) > \delta\}.$$

By Lemma 5.1 (with $\epsilon = \delta$) the map pr in (5) is bijective. As a projection map, pr is continuous. We prove that pr^{-1} is continuous.

Consider $\mathbf{b}_1 \in \mathcal{B}(L, w)$. Let c_1 be the unique element of L with $(\mathbf{b}_1, c_1) \in \big(\mathcal{B}(L, w) \times \mathcal{C}(L, w)\big) \cap V(L)$. Let $\epsilon \in w(L^\times)$ with $\delta \leq \epsilon$. By Lemma 10.1, $w(\mathbf{b}_1, c_1) \geq 0$ and $w(f'(\mathbf{b}_1, c_1)) = \delta$. Let $\mathbf{b} \in \mathcal{B}(L)$ with $w(\mathbf{b} - \mathbf{b}_1) > \delta + \epsilon$. Then the unique element $c \in L$ which Lemma 10.1 (with $(\mathbf{b}_1, c_1)$ replacing $(\mathbf{b}_0, c_0)$) gives satisfies $f(\mathbf{b}, c) = 0$ and $w(c - c_1) > \epsilon$. In particular, $c \in \mathcal{C}(L, w)$, $\mathrm{pr}(\mathbf{b}, c) = \mathbf{b}$, and $w\big((\mathbf{b}, c) - (\mathbf{b}_1, c_1)\big) > \epsilon$, as desired. □

PROPOSITION 10.3. *Let $\varphi: V \to W$ be a morphism of absolutely irreducible varieties over K, $v \in \mathrm{Val}(K)$, $\mathbf{a} \in V_{\mathrm{simp}}(K)$, and $\mathbf{b} \in W_{\mathrm{simp}}(K)$. Suppose φ is étale at $\mathbf{a}$ and $\varphi(\mathbf{a}) = \mathbf{b}$. Then v has an open neighborhood $\mathbf{B}_v$ in $\mathrm{Val}(K)$, $\mathbf{a}$ has an open neighborhood $\mathcal{V}_v$ in $\mathrm{Set}(K, V, \mathbf{B}_v)$, and $\mathbf{b}$ has an open neighborhood $\mathcal{W}_v$ in $\mathrm{Set}(K, W, \mathbf{B}_v)$ satisfying this: For each $(L, w) \in \mathrm{Hensel}(K, \mathbf{B}_v)$ the map $\varphi: \mathcal{V}_v(L, w) \to \mathcal{W}_v(L, w)$ is a w-homeomorphism.*

PROOF. Let $r = \dim(W) = \dim(V)$.

PART A. *Suppose $W = \mathbb{A}^r_K$.* By [Ray, p. 60], φ is locally standard étale. That is, there are a Zariski K-open neighborhood A of $\mathbf{b}$ in $\mathbb{A}^r_K$, a Zariski K-open affine neighborhood V_0 of $\mathbf{a}$ in V, a polynomial $f \in K[T_1, \ldots, T_r, X]$ which is monic in X (and absolutely irreducible), an element $c \in K$, and an isomorphism $\theta: V_0 \to (A \times \mathbb{A}^1_K) \cap V(f)$ over K with $f(\mathbf{b}, c) = 0$, $\frac{\partial f}{\partial X}(\mathbf{b}, c) \neq 0$, $\theta(\mathbf{a}) = (\mathbf{b}, c)$, $\varphi(V_0) = A$, and $\mathrm{pr} \circ \theta = \varphi$. Multiply the point $(\mathbf{b}, c)$ with an appropriate element u of $K^\times$ and the coefficients of f with powers of u and replace θ by $\theta \circ \mu_u$, where $\mu_u(\mathbf{x}, y) = (u\mathbf{x}, uy)$, to assume $v(f) \geq 0$ and $v(\mathbf{b}, c) \geq 0$.

Lemma 10.2 gives an open neighborhood $\mathbf{B}$ of v in $\mathrm{Val}(K)$, an open neighborhood $\mathcal{B}$ of $\mathbf{b}$ in $\mathrm{Set}(K, \mathbb{A}^r_K, \mathbf{B})$, an open neighborhood $\mathcal{C}$ of c in $\mathrm{Set}(K, \mathbb{A}^1_K, \mathbf{B})$ satisfying this:

(6) For each $(L, w) \in \mathrm{Hensel}(K, \mathbf{B})$ the projection $\mathrm{pr}: (\mathcal{B}(L, w) \times \mathcal{C}(L, w)) \cap V(f)(L) \to \mathcal{B}(L, w)$ is a w-homeomorphism.

Replace $\mathcal{B}$ by $\mathcal{B} \cap A$, if necessary, to assume $\mathcal{B} \subseteq A$.

Put $\mathcal{V} = \theta^{-1}\big((\mathcal{B} \times \mathcal{C}) \cap V(f)\big)$. By Definition 9.2, $\mathcal{V}$ is an open neighborhood of $\mathbf{a}$ in $\mathrm{Set}(K, V, \mathbf{B})$. Also, for each $(L, w) \in \mathrm{Hensel}(K, \mathrm{Val}(K))$ the map $\theta: \mathcal{V}(L, w) \to (\mathcal{B}(L, w) \times \mathcal{C}(L, w)) \cap V(f)(L)$ is a w-homeomorphism. If, in addition, $w|_K \in \mathbf{B}$, (6) implies that the map $\varphi: \mathcal{V}(L, w) \to \mathcal{B}(L, w)$ is a w-homeomorphism.

PART B. *The general case.* Since $\mathbf{b}$ is simple on W, the maximal ideal $\mathfrak{m}_{W,\mathbf{b}}$ of the local ring of W has r generators $t_1, \ldots, t_r$, $\tau = (t_1, \ldots, t_r)$ is an étale map of W into $\mathbb{A}^r_K$ at $\mathbf{b}$ and $\tau(\mathbf{b}) = \mathbf{o} = (0, \ldots, 0)$ [Mum, p. 255, Thm. 1]. Part A gives an open neighborhood $\mathbf{B}_1$ of v in $\mathrm{Val}(K)$, an open neighborhood $\mathcal{W}_1$ of $\mathbf{b}$ in $\mathrm{Set}(K, W, \mathbf{B}_1)$, and an open neighborhood $\mathcal{A}_1$ of $\mathbf{o}$ in $\mathrm{Set}(K, \mathbb{A}_K^r, \mathbf{B}_1)$ satisfying this: For each $(L, w) \in \mathrm{Hensel}(K, \mathbf{B}_1)$ the map $\tau: \mathcal{W}_1(L, w) \to \mathcal{A}_1(L, w)$ is a w-homeomorphism.

By [Hrt, p. 268, Prop. 10.1(b)], $\tau \circ \varphi$ is an étale morphism of V into $\mathbb{A}_K^r$ at $\mathbf{a}$. Part A gives an open neighborhood $\mathbf{B}_2$ of v in $\mathrm{Val}(K)$, an open neighborhood $\mathcal{V}_2$ of $\mathbf{a}$ in $\mathrm{Set}(K, V)$, and an open neighborhood $\mathcal{A}_2$ of $\mathbf{o}$ in $\mathrm{Set}(K, \mathbb{A}_K^r, \mathbf{B}_2)$ satisfying this: For all $(L, w) \in \mathrm{Hensel}(K, \mathbf{B}_2)$ the map $\tau \circ \varphi: \mathcal{V}_2(L, w) \to \mathcal{A}_2(L, w)$ is a w-homeomorphism.

Let $\mathbf{B} = \mathbf{B}_1 \cap \mathbf{B}_2$, $\mathcal{A} = \mathcal{A}_1 \cap \mathcal{A}_2$, $\mathcal{W} = \mathcal{W}_1 \cap \tau^{-1}(\mathcal{A}_2)$, and $\mathcal{V} = \varphi^{-1}(\mathcal{W})$. Then $\mathbf{B}, \mathcal{V}, \mathcal{W}$ satisfy the requirements of the lemma. □

COROLLARY 10.4. *Let $\varphi: V \to W$ be a morphism of absolutely irreducible varieties over K, $\mathbf{a} \in V_{\mathrm{simp}}(K)$, $\mathbf{b} \in W_{\mathrm{simp}}(K)$, and $\mathbf{B}$ a closed subset of* $\mathrm{Val}(K)$. *Suppose φ is étale at $\mathbf{a}$ and $\varphi(\mathbf{a}) = \mathbf{b}$. Then there is a partition $\mathbf{B} = \dot\bigcup_{i=1}^n \mathbf{B}_i$ with $\mathbf{B}_i$ closed, $\mathbf{a}$ has an open neighborhood $\mathcal{V}_i$ in $\mathrm{Set}(K, V, \mathbf{B}_i)$, and $\mathbf{b}$ has an open neighborhood $\mathcal{W}_i$ in $\mathrm{Set}(K, W, \mathbf{B}_i)$, $i = 1, \ldots, n$, satisfying this: For all i, $w \in \mathbf{B}_i$, and $(L, w) \in \mathrm{Hensel}(K, w)$ the map $\varphi: \mathcal{V}_i(L, w) \to \mathcal{W}_i(L, w)$ is a w-homeomorphism.*

PROOF. For each $v \in \mathbf{B}$ let $\mathbf{B}_v$, $\mathcal{V}_v$, and $\mathcal{W}_v$ be as in Proposition 10.3. Choose an open-closed subset $\mathbf{B}'_v$ of $\mathrm{Val}(K)$ with $v \in \mathbf{B}'_v \subseteq \mathbf{B}_v$. Then, the collection of all $\mathbf{B}'_v$ is an open covering of $\mathbf{B}$. Since $\mathbf{B}$ is closed in $\mathrm{Val}(K)$ and $\mathrm{Val}(K)$ is compact (Proposition 8.2), $\mathbf{B}$ is compact. Thus there are $v_1, \ldots, v_n \in \mathbf{B}$ such that $\mathbf{B} = \bigcup_{i=1}^n \mathbf{B}'_{v_i}$. Let $\mathbf{B}_i = B'_{v_i} \smallsetminus B'_{v_1} \cup \cdots \cup B'_{v_{i-1}}$, $\mathcal{V}_i = \mathcal{V}_{v_i}$, and $\mathcal{W}_i = \mathcal{W}_{v_i}$. They satisfy the conclusion of the corollary. □

11. Field-Valuation Structures

We extend field structures to "field-valuation structures" by equipping each local field with a valuation.

A **field-valuation structure** is a structure $\mathbf{K} = (K, X, K_x, v_x)_{x \in X}$ satisfying the following conditions:

(1a) $(K, X, K_x)_{x \in X}$ is a field structure. Thus, for each finite separable extension L of K the set $X_L = \{x \in X \mid L \subseteq K_x\}$ is open in X.

(1b) v_x is a valuation of K_x satisfying $v_{x^\sigma} = v_x^\sigma$ for all $x \in X$ and $\sigma \in \mathrm{Gal}(K)$. Here $v_x^\sigma(u^\sigma) = v_x(u)$ for each $u \in K_x$.

(1c) For each finite separable extension L of K define a map $\nu_L: X_L \to \mathrm{Val}(L)$ by $\nu_L(x) = v_x|_L$. Then ν_L is continuous.

The **absolute Galois structure** associated with $\mathbf{K}$ is the same associated with the underlying field structure, namely $\mathrm{Gal}(\mathbf{K}) = (\mathrm{Gal}(K), X, \mathrm{Gal}(K_x))_{x \in X}$. We call **$\mathbf{K}$ proper** if $\mathrm{Gal}(\mathbf{K})$ is proper. We call **$\mathbf{K}$ Henselian** if (K_x, v_x) is Henselian for each $x \in X$.

LEMMA 11.1. *Let $\mathbf{K} = (K, X, K_x, v_x)_{x \in X}$ be a field-valuation structure.*

(a) *Let K' be a separable algebraic extension of K and X' a closed subset of X. Suppose X' is closed under the action of $\mathrm{Gal}(K')$ and $K' \subseteq K_x$ for each $x \in X'$. Then $\mathbf{K}' = (K', X', K_x, v_x)_{x \in X'}$ is a field-valuation structure.*

(b) *For each $x \in X$ let $v_{x,\mathrm{ins}}$ be the unique extension of v_x to $K_{x,\mathrm{ins}}$. Then $\mathbf{K}_{\mathrm{ins}} = (K_{\mathrm{ins}}, X, K_{x,\mathrm{ins}}, v_{x,\mathrm{ins}})$ is a field-valuation structure. Moreover, there is an isomorphism $\mathrm{res}: \mathrm{Gal}(\mathbf{K}_{\mathrm{ins}}) \to \mathrm{Gal}(\mathbf{K})$ of group structures.*

Proof of (a). By Remark 2.6, $(K', X, K_x)_{x\in X'}$ is a field structure. It remains to prove that $\nu_{L'}: X'_{L'} \to \mathrm{Val}(L')$ is continuous for each finite separable extension L' of K'. It suffices to consider $u \in L'$ and to prove that each of the sets $Y = \{x \in X'_{L'} \mid v_x(u) > 0\}$ and $Y' = \{x \in X'_{L'} \mid v_x(u) \geq 0\}$ is open in X'. To this end choose a finite separable extension L of K containing u with $L' = K'L$. Then $Y = X' \cap \{x \in X_L \mid v_x(u) > 0\}$, so Y is open by (1c). Similarly, Y' is open.

PROOF OF (b). It suffices to consider the case when $p = \mathrm{char}(K) > 0$. Let L' be a finite extension of K_{ins} and $u \in L'$. Put $L = K_s \cap L'$. Then $L_{\mathrm{ins}} = L'$ and there is a power q of p with $u^q \in L$. Thus, $\{x \in X \mid L' \subseteq K_{x,\mathrm{ins}}\} = \{x \in X \mid L \subseteq K_x\}$ is open. Also, $v_{x,\mathrm{ins}}(u) = \frac{1}{q} v_x(u^q)$. This implies, $\mathbf{K}_{\mathrm{ins}}$ is a field-valuation structure.□

When all (K_x, v_x) are Henselian, we may replace Condition (1c) by a more convenient one:

LEMMA 11.2. *Let $(K, X, K_x)_{x\in X}$ be a field structure. For each $x \in X$ let v_x be a Henselian valuation on K_x such that $v_{x^\sigma} = v_x^\sigma$ for all $x \in X$ and $\sigma \in \mathrm{Gal}(K)$. Extend each v_x to K_s in the unique possible way. Then $(K, X, K_x, v_x)_{x\in X}$ is a field-valuation structure if and only if*

(2) *the map $\nu: X \to \mathrm{Val}(K_s)$ defined by $x \mapsto v_x$ is continuous.*

PROOF. By the uniqueness of the extension of v_x from K_x to K_s, the equality $v_{x^\sigma} = v_x^\sigma$ holds in $\mathrm{Val}(K_s)$ for all $x \in X$ and $\sigma \in \mathrm{Gal}(K)$.

FIELD-VALUATION STRUCTURE IMPLIES (2). Let $u \in K_s^\times$ and let $x \in X$. We have to show that if $v_x(u) > 0$ (resp. $v_x(u) \geq 0$) and $x' \in X$ is sufficiently close to x, then $v_{x'}(u) > 0$ (resp. $v_{x'}(u) \geq 0$).

Let $f(X) = X^n + a_{n-1}X^{n-1} + \cdots + a_0$ be the irreducible polynomial of u over K_x. Then

$$u = -a_{n-1} - a_{n-2}u^{-1} - \cdots - a_0(u^{-1})^{n-1}. \tag{3}$$

Let $u_1, \ldots, u_n$ be the roots of f in K_s. Since v_x uniquely extends to K_s, we have

$$v_x(u_1) = \cdots = v_x(u_n) = v_x(u). \tag{4}$$

Let N/K be a finite separable extension of K containing $u_1, \ldots, u_n$. Put $L = N \cap K_x$. Then $a_0, \ldots, a_{n-1} \in L$. We distinguish between two cases.

(a) Suppose $v_x(u) > 0$. We have $L \subseteq K_x$. Since $a_0, \ldots, a_{n-1}$ are the elementary symmetric functions in $u_1, \ldots, u_n$, (4) implies that $v_x(a_0), \ldots, v_x(a_{n-1}) > 0$. Hence, by (1), if $x' \in X$ is sufficiently close to x, then $L \subseteq K_{x'}$ and

$$v_{x'}(a_0), \ldots, v_{x'}(a_{n-1}) > 0.$$

It follows $v_{x'}(u) > 0$. Indeed, if $v_{x'}(u) \leq 0$, then $v_{x'}(u^{-1}) \geq 0$, hence, by (3), $v_{x'}(u) > 0$, a contradiction.

(b) Suppose $v_x(u) \geq 0$. We have $L \subseteq K_x$. By (4), $v_x(a_0), \ldots, v_x(a_{n-1}) \geq 0$. Hence, by (1), if $x' \in X$ is sufficiently close to x, then $L \subseteq K_{x'}$ and

$$v_{x'}(a_0), \ldots, v_{x'}(a_{n-1}) \geq 0.$$

It follows that $v_{x'}(u) \geq 0$. Indeed, if $v_{x'}(u) < 0$, then $v_{x'}(u^{-1}) > 0$, and by(3), $v_{x'}(u) \geq 0$, a contradiction.

(2) IMPLIES FIELD-VALUATION STRUCTURE. Let L be a finite separable extension of K. Then, $\nu: X \to \mathrm{Val}(K_s)$ and $\mathrm{res}: \mathrm{Val}(K_s) \to \mathrm{Val}(L)$ are continuous (Lemma 8.4). Hence, $\nu_L = \mathrm{res} \circ \nu|_{X_L}$ is continuous. □

12. Block Approximation

Let $\mathbf{K} = (K, X, K_x, v_x)_{x\in X}$ be a field valuation structure. Put $\mathcal{K} = \{K_x \mid x \in X\}$. Suppose K is P$\mathcal{K}$C, X has only finitely many $\mathrm{Gal}(K)$-orbits, and the restriction of the corresponding valuations to K are independent. Using the local homeomorphism theorem [GPR, Thm. 9.4] for varieties over Henselian fields and the weak approximation theorem, Proposition 3.2 of [HaJ3] proves that K is unirationally closed. In the general case, when X has possibly infinitely many $\mathrm{Gal}(K)$-orbits, the block approximation condition substitutes all three conditions. It says roughly that finitely many algebraic points of a variety V over K, each associated with an open-closed subset of X (a "block") can be simultaneously approximated within the block by a single K-rational point of V. Here is the precise definition:

DEFINITION 12.1. *Block approximation condition.* A **block approximation problem** for a field-valuation structure $\mathbf{K} = (K, X, K_x, v_x)_{x\in X}$ is a data $(V, X_i, L_i, \mathbf{a}_i, c_i)_{i\in I_0}$ satisfying this:

(1a) $(\mathrm{Gal}(L_i), X_i)_{i\in I_0}$ is a special partition of $\mathrm{Gal}(\mathbf{K})$.

(1b) V is a smooth affine variety defined over K.

(1c) $\mathbf{a}_i \in V(L_i)$.

(1d) $c_i \in K^\times$.

An analogous condition where valuations are replaced by orderings appears in [Pre, p. 354] and [FHV, Prop. 1.2].

A **solution** of the problem is a point $\mathbf{a} \in V(K)$ with $v_x(\mathbf{a} - \mathbf{a}_i) > v_x(c_i)$ for all $i \in I_0$ and $x \in X_i$. We say $\mathbf{K}$ satisfies the **block approximation condition** if each block approximation problem for $\mathbf{K}$ is solvable.

Note that we could reformulate the block approximation condition by dropping the condition on V to be smooth and demanding instead $\mathbf{a}_i$ to be smooth on V. □

The block approximation condition has several interesting consequences.

DEFINITION 12.2. *Pseudo-$\mathcal{K}$-closed fields.* Let K be a field and $\mathcal{K}$ a set of field extensions of K. We say K is **P$\mathcal{K}$C** if this holds: Every smooth absolutely irreducible variety V defined over K with a K'-rational point for each $K' \in \mathcal{K}$ has a K-rational point. □

PROPOSITION 12.3. *Let $\mathbf{K} = (K, X, K_x, v_x)_{x\in X}$ be a Henselian field-valuation structure satisfying the block approximation condition.*

(a) *Put $\mathcal{K} = \{K_x \mid x \in X\}$. Then K is P$\mathcal{K}$C.*

(b) *Let $x_1, \ldots, x_n \in X$ lie in distinct* $\mathrm{Gal}(K)$*-orbits. Then $v_{x_1}|_K, \ldots, v_{x_n}|_K$ satisfies the weak approximation theorem.*

(c) *Suppose $x, y \in X$ lie in distinct* $\mathrm{Gal}(K)$*-orbits. Then $v_x|_K$ and $v_y|_K$ are independent.*

(d) *Suppose X has more than one* $\mathrm{Gal}(K)$*-orbit. Then the trivial valuation is not in $\nu_K(X)$.*

(e) *For each $x \in X$, K is v_x-dense in K_x; and*

(f) *(K_x, v_x) is a Henselian closure of $(K, v_x|_K)$.*

(g) *Suppose $K_x \neq K_s$. Then* $\mathrm{Aut}(K_x/K) = 1$.

PROOF OF (a). Let V be a smooth absolutely irreducible variety defined over K with a point $\mathbf{a}_x \in V(K_x)$ for each $x \in X$. Then $\mathrm{Gal}(K(\mathbf{a}_x))$ is an open subgroup of $\mathrm{Gal}(K)$ containing $\mathrm{Gal}(K_x)$. Lemma 3.6 gives a special partition $(\mathrm{Gal}(K(\mathbf{a}_{x_i})), X_i)_{i\in I_0}$ with $x_i \in X_i$ for each $i \in I_0$. Thus, $(V, X_i, K(\mathbf{a}_{x_i}), \mathbf{a}_{x_i}, 1)_{i\in I_0}$ is a block approximation problem for $\mathbf{K}$. Our assumption gives a point $\mathbf{a} \in V(K)$. It follows, K is P$\mathcal{K}$C.

PROOF OF (b). Put $v_i = v_{x_i}|_K$, $i = 1, \ldots, n$. Let a_i, c_i be elements of K with $c_i \neq 0$. Since $X/\mathrm{Gal}(K)$ is profinite, there are open-closed distinct $\mathrm{Gal}(K)$-invariant subsets $X_1, \ldots, X_n$ of X with $x_i \in X_i$, $i = 1, \ldots, n$. Let $I = \{0, 1, \ldots, n\}$, $X_0 = X \smallsetminus X_1 \cup \cdots \cup X_n$, $a_0 = 0$, and $c_0 = 1$. Then $(\mathbb{A}^1_K, X_i, K, a_i, c_i)_{i\in I_0}$ is a block approximation problem for $\mathbf{K}$.

By assumption, there is an $a \in K$ with $v_i(a - a_i) > v_i(c_i)$, $i = 1, \ldots, n$. It follows, $v_1, \ldots, v_n$ satisfy the weak approximation theorem.

PROOF OF (c). Use (b).

PROOF OF (d). Assume $v_0 = v_x|_K$ is trivial for some $x \in X$. Choose $y \in X$ outside the $\mathrm{Gal}(K)$-orbit of x. By (c), $v_1 = v_y|_K$ is nontrivial. Hence, there is an $a_1 \in K$ with $v_1(a_1) < 0$. Statement (b) gives $a \in K$ with $v_0(a - a_1) > 0$ and $v_1(a) > 0$. By the first inequality, $a = a_1$. Hence, by the second inequality, $v_1(a_1) > 0$, in contradiction to the choice of a_1.

PROOF OF (e) Let $x \in X$, $a_1 \in K_x$, and $c_1 \in K^\times$. We have to find $a \in K$ satisfying $v_x(a - a_1) > v_x(c_1)$.

By assumption $K(a_1) \leq K_x$ and the stabilizer of x is contained in $\mathrm{Gal}(K_x)$. Therefore, Lemma 3.4 gives an open-closed neighborhood X_1 of x in X which is invariant under $\mathrm{Gal}(K(a_1))$ such that $X_1^{\mathrm{Gal}(K)} = \mathop{\dot\bigcup}_{\rho\in R_1} X_1^\rho$ for each $R_1 \subseteq \mathrm{Gal}(K)$ with $\mathrm{Gal}(K) = \mathop{\dot\bigcup}_{\rho\in R_1} \mathrm{Gal}(K(a_1))\rho$. Thus, $\mathrm{Gal}(K(a_1)) = \{\sigma \in \mathrm{Gal}(K) \mid X_1^\sigma = X_1\}$. Put $L_1 = K(a_1)$.

Let $I_0 = \{0, 1\}$, $X_0 = X \smallsetminus X_1^{\mathrm{Gal}(K)}$, $a_0 = 0$, $c_0 = 1$, and $L_0 = K$. Then $(\mathbb{A}^1_K, X_i, L_i, a_i, c_i)_{i=0,1}$ is a block approximation problem for $\mathbf{K}$.

By assumption, there is an $a \in K$ with $v_x(a - a_1) > v_x(c_1)$, as desired.

PROOF OF (f). By assumption, (K_x, v_x) is Henselian. Choose a Henselian closure (K', v_x) of $(K, v_x|_K)$ in (K_x, v_x). Consider $a \in K_x$. Let $a_1, \ldots, a_n$ be the conjugates of a over K'. By (e) there is a $b \in K$ with $v_x(b - a) > \max_{i\neq j} v_x(a_i - a_j)$. Hence, by Krasner's Lemma [Jar, Lemma 12.1], $K'(a) \subseteq K'(b) = K'$. Therefore, $K_x = K'$.

PROOF OF (g). Let $\sigma \in \mathrm{Aut}(K_x/K)$. Then both v_x and v_x^σ are Henselian valuations of K_x. Therefore, K_x has a nontrivial valuation w which is coarser than both v_x and v_x^σ [Jar, Lemma 13.2]. In particular, the v_x-topology of K coincides with the w-topology of K [Jar, Lemma 3.2]. Hence, by (e), K is w-dense in K_x.

Assume there exists $b \in K_x$ with $b \neq b^\sigma$. Then there exists $c \in K^\times$ with $v(c) > v(b - b^{\sigma^{-1}})$ and there exists $a \in K$ with $w(a - b) > w(c)$. Since w is coarser than both v and v^σ, we have $v(a - b) > v(c)$ and $v^\sigma(a - b) > v(c)$. Hence, $v(a - b^{\sigma^{-1}}) > v(c)$. Therefore, $v(b - b^{\sigma^{-1}}) > v(c)$, in contradiction to the choice of c. □

PROPOSITION 12.4. *Let* $\mathbf{K}$ *be a Henselian field-valuation structure that satisfies the block approximation condition. Then* $\mathbf{K}$ *is unirationally closed.*

PROOF. Consider a unirational arithmetical problem

$$\Phi = (V, X_i, L_i, \pi_i: U_i \to V \times_K L_i)_{i \in I_0}$$

for $\mathbf{K}$ as in Definition 6.1. Let $X' = \bigcup_{i \in I_0} X_i$. We find a solution $(\mathbf{a}, \mathbf{b}_x)_{x \in X'}$ of Φ.

To this end consider $i \in I_0$. Since ν_{L_i} is continuous, $\mathbf{B}_i = \nu_{L_i}(X_i)$ is a closed subset of $\mathrm{Val}(L_i)$. For each $x \in X_i$ put $v_{x,i} = \nu_{L_i}(x)$. Then $(K, v_x|_K) \subseteq (L_i, v_{x,i}) \subseteq (K_x, v_x)$.

Since U_i is birationally equivalent to $\mathbb{A}^r_{L_i}$, there exists $\mathbf{a}_i \in U_i(L_i)$. Then $\mathbf{b}_i = \pi_i(\mathbf{a}_i) \in V(L_i)$. By definition, π_i is étale at $\mathbf{a}_i$ (see (3d) of Section 6). Thus, Corollary 10.4 (with L_i, $\pi_i: U_i \to V \times_K L_i$ replacing K, $\varphi: V \to W$) gives a partition $\mathbf{B}_i = \dot{\bigcup}_{j \in J_i} \mathbf{B}_{ij}$ with $\mathbf{B}_{ij}$ closed in $\mathrm{Val}(L_i)$, an open neighborhood $\mathcal{U}_{ij}$ of $\mathbf{a}_i$ in $\mathrm{Set}(L_i, U_i, \mathbf{B}_{ij})$, and an open neighborhood $\mathcal{V}_{ij}$ of $\mathbf{b}_i$ in $\mathrm{Set}(L_i, V \times_K L_i, \mathbf{B}_{ij})$, $j \in J_i$ satisfying this:

(2) For all $j \in J_i$ and $x \in X_i$ with $v_{x,i} \in \mathbf{B}_{ij}$ the map $\pi_i: \mathcal{U}_{ij}(K_x) \to \mathcal{V}_{ij}(K_x)$ is a v_x-homeomorphism.

For all $i \in I_0$ and $j \in J_i$ Lemma 9.4 gives a partition $\mathbf{B}_{ij} = \dot{\bigcup}_{l \in \Lambda_{ij}} \mathbf{B}_{ijl}$ with Λ_{ij} finite, $\mathbf{B}_{ijl}$ closed, and $c_{ijl} \in L_i^\times$, $l \in \Lambda_{ij}$, such that $\mathcal{B}_{\mathbf{b}_i, c_{ijl}, \mathbf{B}_{ijl}}(M, w) \subseteq \mathcal{V}_{ij}(M, w)$ for each $(M, w) \in \mathrm{Hensel}(L_i, \mathbf{B}_{ijl})$, $l \in \Lambda_{ij}$. For all $l \in \Lambda_{ij}$ put $L_{ijl} = L_i$, $X_{ijl} = \nu_{L_i}^{-1}(\mathbf{B}_{ijl})$, and $\mathbf{b}_{ijl} = \mathbf{b}_i$. Then X_{ijl} is a closed subset of X_i, $X_i = \dot{\bigcup}_{j \in J_i} \dot{\bigcup}_{l \in \Lambda_{ij}} X_{ijl}$, and

$$\{\mathbf{b} \in V(K_x) \mid v_x(\mathbf{b} - \mathbf{b}_i) > v_x(c_{ijl})\} \subseteq \mathcal{V}_{ij}(K_x) \tag{3}$$

for all $x \in X_{ijl}$ and $l \in \Lambda_{ij}$.

Since X_i is open-closed in X, so are X_{ijl}. If $\sigma \in \mathrm{Gal}(L_i)$, then $X_{ijl}^\sigma = X_{ijl}$. Indeed, let $x \in X_{ijl}$. Then $\nu_{L_i}(x^\sigma) = v_{x^\sigma}|_{L_i} = v_x^\sigma|_{L_i} = v_x|_{L_i} \in \mathbf{B}_{ijl}$, so $x^\sigma \in X_{ijl}$. If $\sigma \in \mathrm{Gal}(K)$, $i, i' \in I_0$, $j \in J_i$, $j' \in J_{i'}$, $l \in \Lambda_{ij}$, $l' \in \Lambda_{i'j'}$, and $X_{ijl}^\sigma \cap X_{i'j'l'} \neq \emptyset$, then $X_i^\sigma \cap X_{i'} \neq \emptyset$, so $i' = i$ and $\sigma \in \mathrm{Gal}(L_i)$. Thus our assumption becomes $X_{ijl} \cap X_{ij'l'} \neq \emptyset$. Therefore, $j = j'$ and $l = l'$. It follows that

$$(V, X_{ijl}, L_{ijl}, \mathbf{a}_{ijl}, c_{ijl})_{i \in I_0,\, j \in J_i,\, l \in \Lambda_{ij}}$$

is a block approximation problem for $\mathbf{K}$.

The block approximation condition gives $\mathbf{b} \in V(K)$ with $v_x(\mathbf{b} - \mathbf{a}_i) > v_x(c_{ijl})$ for all $i \in I_0$, $j \in J_i$, $l \in \Lambda_{ij}$, and $x \in X_{ijl}$. By (3), $\mathbf{b} \in \mathcal{V}_{ij}(K_x)$. By (2), there is an $\mathbf{a}_x \in \mathcal{U}_{ij}(K_x)$ with $\pi_i(\mathbf{a}_x) = \mathbf{b}$. In particular, $\mathbf{a}_x \in U_i(K_x)$. Thus, $(\mathbf{b}, \mathbf{a}_x)_{x \in X'}$ is a solution of Φ. □

THEOREM 12.5. *Let* $\mathbf{K} = (K, X, K_x, v_x)_{x \in X}$ *be a proper Henselian field-valuation structure. Suppose* $\mathbf{K}$ *satisfies the block approximation condition. Then* $\mathrm{Gal}(\mathbf{K})$ *is a projective group structure.*

PROOF. By Proposition 12.4, $\mathbf{K}$ is unirationally closed. Since $\mathrm{Gal}(\mathbf{K})$ is a proper group structure, $S_x = \mathrm{Gal}(K_x)$ for each $x \in X$ (Remark 2.1). Hence, by Proposition 6.4, $\mathrm{Gal}(\mathbf{K})$ is projective. □

This completes the proof of Part (a) of the Main Theorem. The rest of the work is devoted to the proof of Part (b) of the Main Theorem.

13. Rigid Henselian Extensions

This section continuous of Section 7. It contains various results about valued fields which are needed in the proof of Part (b) of the Main Theorem.

For a field extension F/K let $\mathrm{Val}(F/K)$ be the space of all valuations of F (including the trivial one) which are trivial on K. Denote the valuation ring of a valuation w of K by O_w, its maximal ideal by M_w, and its residue field by $\bar{K}_w$. Another valuation v of K is said to be **finer than** w if $O_v \subseteq O_w$, equivalently if $M_w \subseteq M_v$. Thus, $w(x) < w(y)$ implies $v(x) < v(y)$ for all $x, y \in K$. Then $E = \bar{K}_w$ has a unique valuation $\bar{v}$ satisfying $\bar{v}(x + M_w) = v(x)$ for $x \in O_w$. In particular, $\bar{K}_v = \bar{E}_{\bar{v}}$. Denote $\bar{v}$ by v/w.

Conversely, given a valuation $\bar{v}$ of E, there is a unique valuation v of K which is finer than w for which $v/w = \bar{v}$ [Jar, §3]. Then the place $\varphi_v \colon K \to \bar{K}_v \cup \{\infty\}$ corresponding to v is the compositum of the place $\varphi_w \colon K \to \bar{K}_w \cup \{\infty\}$ and the place $\varphi_{\bar{v}} \colon \bar{K}_w \to \bar{K}_v \cup \{\infty\}$. We write $v = \bar{v} \cdot w$.

Lemma 13.1. *Let K be a field, $\tilde{K}$ its algebraic closure, and T a set of indeterminates with* $\mathrm{card}(T) \geq \mathrm{card}(\tilde{K})$. *Put $F = K(T)$. Then, for each algebraic extension L of K there exists $v \in \mathrm{Val}(F/K)$ with $\bar{F}_v = L$.*

Proof. Put $m = \mathrm{card}(T)$. Choose a well ordered transfinite sequence $(a_\alpha)_{\alpha<m}$ which generates L over K. Well-order T as $(t_\alpha)_{\alpha<m}$. For each $\beta \leq m$ let $F_\beta = K(t_\alpha \mid \alpha \leq \beta)$ and $L_\beta = L(a_\alpha \mid \alpha \leq \beta)$.

Consider $\gamma \leq m$. Inductively suppose for each $\beta < \gamma$ there is a $v_\beta \in \mathrm{Val}(F_\beta/K)$ with $\bar{F}_\beta = L_\beta$ such that $v_{\beta'}$ extends v_β whenever $\beta \leq \beta'$.

If γ is a limit cardinal, then the union of all v_β is a valuation v_γ of F_γ with residue field L_γ. Otherwise, $\gamma = \beta + 1$, $F_\gamma = F_\beta(t_\gamma)$, and t_γ is transcendental over F_β. Extend v_β to a valuation v' of F_γ with residue field $L_\beta(t_\gamma)$ with t_γ being its own residue [Bou, Chap. VI, §10.1, Lemma 1, p. 434]. Let w be the L_β-valuation of $L_\beta(t_\gamma)$ with $\bar{t}_\gamma = a_\gamma$ and $\overline{L_\beta(t_\gamma)} = L_\beta(a_\gamma) = L_\gamma$. Then $\varphi_w \circ \varphi_{v'}$ extends φ_{v_β}. Hence, $v_\gamma = w \cdot v'$ extends v_β and has L_γ as residue field. This completes the induction.

The valuation $v = v_m$ of F is trivial on K and satisfies $\bar{F}_v = L$. □

Lemma 13.2. *Consider a perfect field K.*

(a) *Let L be an extension of K and $v \in \mathrm{Val}(L/K)$. Suppose (L, v) is Henselian, $\bar{L}_v$ is an algebraic extension of K, and* $\mathrm{res} \colon \mathrm{Gal}(L) \to \mathrm{Gal}(K)$ *is an isomorphism. Then, $\bar{L}_v = K$.*

(b) *Let L be a rigid Henselian extension of K (Definition 7.6) and L' a separable algebraic extension of L. Then L' is a rigid Henselian extension of $L' \cap \tilde{K}$.*

(c) *Suppose L/K and M/L are rigid Henselian extensions. Then so is M/K.*

(d) *Let K be a field and I a totally ordered set. For each $i \in I$ let (L_i, v_i) be a rigid Henselian extension of K. Suppose $(L_i, v_i) \subseteq (L_j, v_j)$ if $i \leq j$. Put $(L, v) = \bigcup_{i \in I}(L_i, v_i)$. Then (L, v) is a rigid Henselian extension of K.*

Proof of (a). By Lemma 7.4(a), reduction modulo v defines an epimorphism $\rho \colon \mathrm{Gal}(L) \to \mathrm{Gal}(\bar{L}_v)$ and $\rho = \mathrm{res}_{L_s/\tilde{K}}$. Hence,

$$\mathrm{Gal}(\bar{L}_v) = \rho(\mathrm{Gal}(L)) = \mathrm{res}_{L_s/\tilde{K}}(\mathrm{Gal}(L)) = \mathrm{Gal}(K).$$

Therefore, $K = \bar{L}_v$.

PROOF OF (b). By definition, L has a valuation v such that (L, v) is Henselian, $\bar{L}_v = K$, and $\text{res}\colon \text{Gal}(L) \to \text{Gal}(K)$ is an isomorphism. Denote the unique extension of v to L' by v. Then, (L', v) is Henselian, $\overline{L'}_v/K$ is algebraic, and $\text{res}\colon \text{Gal}(L') \to \text{Gal}(L' \cap \tilde{K})$ is an isomorphism. By (a), $\overline{L'}_v = L' \cap \tilde{K}$. Therefore, (L', v) is a rigid Henselian extension of $L' \cap \tilde{K}$.

PROOF OF (c). By assumption, L admits a valuation v and M admits a valuation w such that (L, v) is a rigid Henselian extension of K and (M, w) is a rigid Henselian extension of L. Let $w' = v \cdot w$. Then (M, w') is Henselian and $\bar{M}_{w'} = K$ [Jar, Prop. 13.1]. Also, $\varphi_{w'}(a) = \varphi_v(\varphi_w(a)) = a$ for each $a \in K$. Hence, w' is trivial on K. Finally, $\text{res}\colon \text{Gal}(M) \to \text{Gal}(L)$ and $\text{res}\colon \text{Gal}(L) \to \text{Gal}(K)$ are isomorphisms. Therefore, $\text{res}\colon \text{Gal}(M) \to \text{Gal}(K)$ is an isomorphism. Consequently, (M, w') is a rigid Henselian extension of K.

PROOF OF (d). Routine check. □

An earlier version of the following result appears on page 24 of [Pop] without a proof.

LEMMA 13.3. *Let K be a field and T a set of indeterminates with* $\text{card}(T) \geq \text{card}(\tilde{K})$. *Put $F = K(T)_{\text{ins}}$. Then, for each perfect algebraic extension L of K there are $v \in \text{Val}(F/K)$ and a Henselian closure (F_v, v) of (F, v) which is a rigid Henselian extension of L.*

PROOF. We may replace K by K_{ins}, if necessary, to assume K is perfect. Write $T = \dot{\bigcup}_{i=1}^{\infty} T_i$ with $\text{card}(T_i) = \text{card}(T)$ for each i. Inductively define $K_0 = K$ and $K_i = K_{i-1}(T_i)_{\text{ins}}$ for $i = 1, 2, 3 \ldots$. Then K_i is perfect and $\text{card}(\tilde{K}_i) = \text{card}(T_i)$, $i = 1, 2, 3, \ldots$. Also, $F = \bigcup_{i=1}^{\infty} K_i$.

Let v_0 be the trivial valuation of K. Put $K_0' = K_0$ and $L_0 = L$. Suppose by induction we have constructed algebraic extensions $K_i' \subseteq L_i$ of K_i and a valuation v_i of L_i satisfying this:

(1a) (K_i', v_i) is a Henselian closure of $(K_i, v_i|_{K_i})$.
(1b) (L_i, v_i) is a rigid Henselian extension of L.
(1c) $L_{i-1} \subseteq K_i'$.
(1d) v_i extends v_{i-1}.

Lemma 13.1 gives a valuation $w \in \text{Val}(K_{i+1}/K_i)$ with residue field L_i. Let $v_{i+1} = v_i \cdot w$. Since L_i/K_i is separable, (K_{i+1}, w) has a Henselian closure E which contains L_i. Since v_{i+1} is finer than w, there is a Henselian closure (K_{i+1}', v_{i+1}) of (K_{i+1}, v_{i+1}) which contains E [Jar, Cor. 14.4], hence L_i. By Proposition 7.4(c), K_{i+1}' has an algebraic extension L_{i+1} such that $\text{res}\colon \text{Gal}(L_{i+1}) \to \text{Gal}(L)$ is an isomorphism. Denote the unique extension of v_{i+1} to L_{i+1} again by v_{i+1}. Then (L_{i+1}, v_{i+1}) is a rigid Henselian extension of L (Lemma 13.2(b)).

Let $F_v = \bigcup_{i=1}^{\infty} K_i'$, $L_\infty = \bigcup_{i=1}^{\infty} L_i$, and $v = \bigcup_{i=1}^{\infty} v_i$. Then v is a valuation of F_v over K, (F_v, v) is a Henselian closure of (F, v), $F_v = L_\infty$, $L \subseteq L_\infty$, and $\text{res}\colon \text{Gal}(L_\infty) \to \text{Gal}(L)$ is an isomorphism. Thus, (F_v, v) is a rigid Henselian extension of L. □

LEMMA 13.4. *Let (K, v) be a valued field and (E, w) a Henselian closure. Suppose $E \neq K_s$ and for each separable algebraic extension $F \neq K_s$ of E the residue field $\bar{F}$ of F under the unique extension of w to F is not separably closed. Then* $\text{Aut}(E/K) = 1$ *and $EE^\sigma = K_s$ for each $\sigma \in \text{Gal}(K) \smallsetminus \text{Gal}(E)$.*

PROOF. By assumption, $\bar{E}$ is not separably closed. Hence, by F. K. Schmidt - Engler, $\mathrm{Aut}(E/K) = 1$ [Jar, Prop. 14.5].

Now consider $\sigma \in \mathrm{Gal}(K)$. Put $E' = E^\sigma$ and $w' = w^\sigma$. Then (E', w') is also a Henselian closure of (K, v). Let $F = EE'$. Denote the unique extension of w (resp. w') to F by w_F (resp. w'_F). Then both w_F and w'_F extend v. We prove: Either $\sigma \in \mathrm{Gal}(E)$ or $F = K_s$.

CASE A. $w_F = w'_F$. Denote the unique extension of w_F to K_s by w_s. It coincides with the unique extension w'_s of w'_F to K_s. In addition, w_s^σ is the unique extension of w' to K_s, so also the unique extension of w'_F to K_s. Thus, $w_s = w'_s = w_s^\sigma$. Therefore, σ belongs to the decomposition group of w_s over E, which is $\mathrm{Gal}(E)$.

CASE B. $w_F \neq w'_F$. By Engler, w_F and w'_F are incomparable [Jar, Prop. 6.6]. Since F is Henselian with respect to both w_F and w'_F, the field $\bar{F}_{w_F}$ is separably closed [Jar, Prop. 13.4]. Hence, by assumption, $F = K_s$. □

14. Projective Group Structures as Absolute Galois Structures

Part (b) of the Main Theorem gives for each proper projective group structure $\mathbf{G} = (G, X, G_x)_{x \in X}$ a proper field-valuation structure $\mathbf{L}$ and isomorphism $\lambda\colon \mathbf{G} \to \mathrm{Gal}(\mathbf{L})$. We call λ a **Galois isomorphism of G**. An obvious necessary condition for the existence of a Galois isomorphism of $\mathbf{G}$ is the existence of a **Galois approximation** of $\mathbf{G}$. This is a rigid epimorphism $\kappa\colon \mathbf{G} \to \mathrm{Gal}(\mathbf{K})$ where $\mathbf{K}$ is a field structure. In this section we generalize [Pop, Thm. 3.4] and "lift" each Galois approximation of $\mathbf{G}$ to an isomorphism $\kappa'\colon \mathbf{G} \to \mathrm{Gal}(\mathbf{K}')$ where $\mathbf{K}'$ is a field structure. Then, in Section 15, we lift κ' further to a Galois isomorphism λ as above.

Here $\kappa'\colon \mathbf{G} \to \mathrm{Gal}(\mathbf{K}')$ is said to **lift** κ if $K \subseteq K'$ and $\mathrm{res}\colon \mathrm{Gal}(K') \to \mathrm{Gal}(K)$ extends to a morphism $\rho\colon \mathrm{Gal}(\mathbf{K}') \to \mathrm{Gal}(\mathbf{K})$ with $\mathrm{res} \circ \kappa' = \kappa$. Then ρ is a rigid epimorphism.

LEMMA 14.1. *Let G be a profinite group, H an open subgroup, K a closed normal subgroup, and $\mathcal{G}$ an étale compact subset of* $\mathrm{Subgr}(G)$. *Suppose $\Gamma \cap K = 1$ for each $\Gamma \in \mathcal{G}$. Then G has an open normal subgroup N with $N \le H$ and $\Gamma N \cap KN = N$ for each $\Gamma \in \mathcal{G}$.*

PROOF. Let $\mathcal{N}$ be the set of open normal subgroups of G containing K. Assume without loss $H \triangleleft G$. Now consider $\Delta \in \mathcal{G}$. Assume, for each $M \in \mathcal{N}$, the closed subset $\Delta \cap M \smallsetminus H$ of G is nonempty. Then, by compactness of G, $\bigcap_{M \in \mathcal{N}} \Delta \cap M \smallsetminus H \neq \emptyset$. On the other hand, $\bigcap_{M \in \mathcal{N}} \Delta \cap M = \Delta \cap \bigcap_{M \in \mathcal{N}} M = \Delta \cap K = 1$. This contradiction gives $M_\Delta \in \mathcal{N}$ with $\Delta \cap M_\Delta \smallsetminus H = \emptyset$. In other words, $\Delta \cap M_\Delta \le H$. It follows that $\Delta(H \cap M_\Delta) \cap M_\Delta \le H$.

Now consider the étale open neighborhood $\mathcal{U}_\Delta = \mathrm{Subgr}\big(\Delta(H \cap M_\Delta)\big) \cap \mathcal{G}$ of Δ in $\mathcal{G}$. For each $\Gamma \in \mathcal{U}_\Delta$ we have $\Gamma \cap M_\Delta \le \Delta(H \cap M_\Delta) \cap M_\Delta \le H$.

Since $\mathcal{G}$ is étale compact, there are $\Delta_1, \dots, \Delta_r \in \mathcal{G}$ with $\mathcal{G} = \bigcup_{i=1}^r \mathcal{U}_{\Delta_i}$. Then $N = H \cap \bigcap_{i=1}^r M_{\Delta_i}$ is the desired open normal subgroup of G. Indeed, let $\Gamma \in \mathcal{G}$. Then $\Gamma \in \mathcal{U}_{\Delta_j}$ for some j. Hence, $\Gamma \cap KN \le \Gamma \cap M_{\Delta_j} \le H$. Thus, $\Gamma \cap KN \le H \cap \bigcap_{i=1}^r M_{\Delta_i} = N$. Therefore, $\Gamma N \cap KN = N$. □

In the following Lemma and its applications we use the relation $A \subset B$ between sets to mean "A is a proper subset of B".

LEMMA 14.2. *Let* $\mathbf{G} = (G, X, G_x)_{x\in X}$ *be a proper projective group structure,* $\kappa\colon \mathbf{G} \to \mathrm{Gal}(\mathbf{K})$ *a Galois approximation, and* G_0 *an open subgroup of* G. *Then* κ *can be lifted to a Galois approximation* $\epsilon\colon \mathbf{G} \to \mathrm{Gal}(\mathbf{E}')$ *with* $K \subset E'$, $\mathrm{Ker}(\epsilon) \le G_0$, *and* $\mathrm{trans.deg}(E'/K) < \infty$.

PROOF. Replacing $\mathbf{K}$ by $\mathbf{K}_{\mathrm{ins}}$ (Lemma 11.1), if necessary, we may assume K is perfect. The rest of the proof has three parts.

PART A. *Replacing* $\mathrm{Gal}(K)$ *by a relative Galois group.* By definition, $G_x \cap \mathrm{Ker}(\kappa) = 1$ for each $x \in X$. Hence, Lemma 14.1 gives an open normal subgroup N of G contained in G_0 with

$$G_xN \cap \mathrm{Ker}(\kappa)N = N \qquad \text{for each } x \in X. \tag{1}$$

Put $B = G/N$ and $A = \mathrm{Gal}(K)/\kappa(N)$. Let $\beta\colon G \to B$ and $\iota\colon \mathrm{Gal}(K) \to A$ be the quotient maps, and $\alpha\colon B \to A$ the epimorphism induced by κ. Then $\alpha \circ \beta = \iota \circ \kappa$. Let $\bar{G} = B \times_A \mathrm{Gal}(K)$. Then let $\bar{\kappa}\colon \bar{G} \to \mathrm{Gal}(K)$ and $\bar{\beta}\colon \bar{G} \to B$ be the coordinate projections. There is a unique morphism $\rho\colon G \to \bar{G}$ with $\bar{\kappa} \circ \rho = \kappa$ and $\bar{\beta} \circ \rho = \beta$.

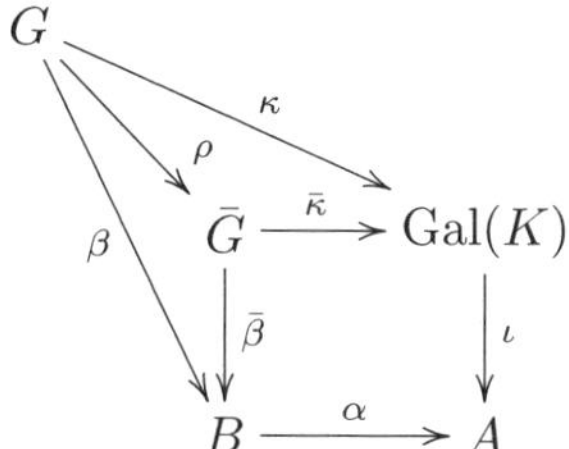

Since $\mathrm{Ker}(\iota \circ \kappa) = N\mathrm{Ker}(\kappa) = \mathrm{Ker}(\beta)\mathrm{Ker}(\kappa)$, we may assume that $\bar{G} = G/N \cap \mathrm{Ker}(\kappa)$ and ρ is the quotient map [FrJ, Section 22.2].

Let L be the fixed field of $\kappa(N)$ in $\tilde{K}$. Identify A with $\mathrm{Gal}(L/K)$ and ι with $\mathrm{res}_{\tilde{K}/L}$. Lemma 6.2 gives a regular extension E of K of transcendence degree equal to $|B|$ (in particular, $E \ne K$) and a finite Galois extension F of E containing L with $B = \mathrm{Gal}(F/E)$ and $\alpha = \mathrm{res}_{F/L}$. Since E/K is regular, $\bar{G} = \mathrm{Gal}(F/E) \times_{\mathrm{Gal}(L/K)} \mathrm{Gal}(K) = \mathrm{Gal}(F\tilde{K}/E)$, $\bar{\beta} = \mathrm{res}_{F\tilde{K}/F}$, and $\bar{\kappa} = \mathrm{res}_{F\tilde{K}/\tilde{K}}$.

Extend $\bar{G}$ to a group structure $\bar{\mathbf{G}} = \mathbf{G}/\mathrm{Ker}(\rho)$ and ρ to the quotient map $\rho\colon \mathbf{G} \to \bar{\mathbf{G}}$. Then $\bar{\kappa}$ extends to a rigid epimorphism $\bar{\kappa}\colon \bar{\mathbf{G}} \to \mathrm{Gal}(\mathbf{K})$ such that $\kappa = \bar{\kappa} \circ \rho$.

PART B. *The cover* $\pi\colon \mathrm{Gal}(\mathbf{E}) \to \bar{\mathbf{G}}$. Write $\bar{\mathbf{G}}$ as $(\bar{G}, Y, \bar{G}_y)_{y\in Y}$. Put $\bar{N} = \rho(N)$. For each $y \in Y$ choose $x \in X$ such that $\rho(x) = y$. Then $\bar{G}_y = \rho(G_x)$ and $\bar{G}_y\bar{N} = \rho(G_xN)$ is an open subgroup of $\bar{G}$ which contains $\bar{G}_y$. Let L_y be the fixed field of $\bar{\kappa}(\bar{G}_y\bar{N}) = \kappa(G_xN)$ in $\tilde{K}$ and F_y the fixed field of $\bar{\beta}(\bar{G}_y\bar{N}) = \beta(G_xN) = G_xN/N$ in F. Then $\kappa(G_xN) = \mathrm{Gal}(L_y)$, $\beta(G_xN) = \mathrm{Gal}(F/F_y)$, and $\bar{G}_y\bar{N} = \mathrm{Gal}(F\tilde{K}/F_y)$. Since $\alpha = \mathrm{res}_{F/L}$ maps $\mathrm{Gal}(F/F_y)$ onto $\mathrm{res}_{\tilde{K}/L}(\mathrm{Gal}(L_y)) = \mathrm{Gal}(L/L_y)$, we have $L_y \subseteq F_y$. Also, $\mathrm{Ker}(\alpha) = \mathrm{Ker}(\kappa)N/N$. Hence, by (1), α is injective on G_xN/N. Thus α maps $\mathrm{Gal}(F/F_y)$ isomorphically onto $\mathrm{Gal}(L/L_y)$. By Lemma 6.2, F_y/L_y is a purely transcendental extension.

Proposition 7.7(c) gives a perfect algebraic extension E_y of F_y which is a rigid Henselian extension of L_y. In particular, $\mathrm{res}_{\tilde{E}/\tilde{K}}\colon \mathrm{Gal}(E_y) \to \mathrm{Gal}(L_y)$ is an isomorphism. Therefore, $\tilde{E} = E_y\tilde{K}$ and $F\tilde{K} = F_yL\tilde{K} = F_y\tilde{K}$. Consequently,

res: $\mathrm{Gal}(E_y) \to \mathrm{Gal}(F\tilde{K}/F_y)$ is an isomorphism.

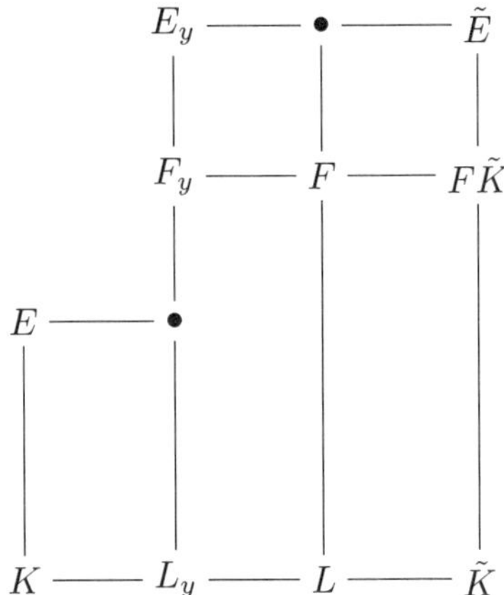

Lemma 3.6 gives a finite subset $\{y_i \mid i \in I_0\}$ of Y and a special partition $(\bar{G}_i, Y_i, R_i)_{i\in I_0}$ of $\bar{\mathbf{G}}$ (Definition 3.5) such that $\bar{G}_i = \bar{G}_{y_i}\bar{N}$ and $y_i \in Y_i$ for each $i \in I_0$. Thus, $\mathrm{res}_{\tilde{E}/F\tilde{K}}\colon \mathrm{Gal}(E_{y_i}) \to \bar{G}_i$ is an isomorphism, $i \in I_0$. Therefore, Lemma 5.1 extends $\mathrm{res}_{\tilde{E}/F\tilde{K}}\colon \mathrm{Gal}(E) \to \bar{G}$ to a cover of group structures. This means there is a field structure $\mathbf{E}$ on E and a cover $\pi\colon \mathrm{Gal}(\mathbf{E}) \to \bar{\mathbf{G}}$.

PART C. *Applying projectivity.* Since $\mathbf{G}$ is projective, there is a morphism $\epsilon\colon \mathbf{G} \to \mathrm{Gal}(\mathbf{E})$ with $\pi \circ \epsilon = \rho$. Let E' be the fixed field of $E(G)$ in $\tilde{E}$. Then E' extends to a field structure $\mathbf{E}'$ such that $\mathrm{Gal}(\mathbf{E}')$ is a sub-group-structure of $\mathrm{Gal}(\mathbf{E})$ and $\epsilon\colon \mathbf{G} \to \mathrm{Gal}(\mathbf{E}')$ is an isomorphism. Since both ρ and π are covers, $\rho\colon G_x \to \bar{G}_{\rho(x)}$ and $\pi\colon \mathrm{Gal}(E'_{\epsilon(x)}) \to \bar{G}_{\rho(x)}$ are isomorphisms, so $\epsilon\colon G_x \to \mathrm{Gal}(E'_{\epsilon(x)})$ is an isomorphism for each $x \in X$. Thus ϵ is a rigid epimorphism, hence ϵ is a Galois approximation of $\mathbf{G}$ which lifts κ.

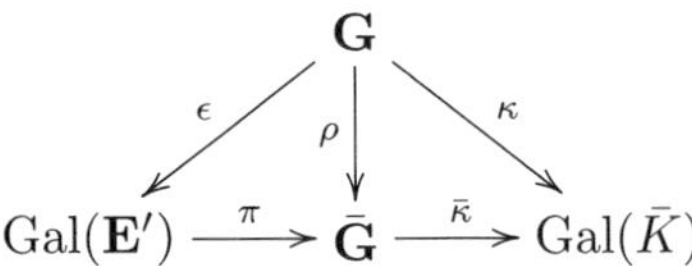

Finally $\mathrm{Ker}(\epsilon) \le \mathrm{Ker}(\rho) \le N \le G_0$ and E' is a proper extension of K. □

PROPOSITION 14.4. *Let $\mathbf{G}$ be a proper projective group structure and $\kappa\colon \mathbf{G} \to \mathrm{Gal}(\mathbf{K})$ a Galois approximation. Then κ can be lifted to a Galois isomorphism $\lambda\colon \mathbf{G} \to \mathrm{Gal}(\mathbf{L})$ with an underlying perfect field.*

PROOF. Let $\{G_\alpha \mid \alpha < m\}$ be a well ordering of all open subgroups of G. By transfinite induction we construct for each $\alpha \le m$ a Galois approximation $\kappa_\alpha\colon \mathbf{G} \to \mathrm{Gal}(\mathbf{K}_\alpha)$ such that $\mathbf{K}_0 = \mathbf{K}$, $\kappa_0 = \kappa$, κ_β lifts κ_α if $\alpha \le \beta \le m$, the underlying field K_α of $\mathbf{K}_\alpha$ is perfect, and $\mathrm{Ker}(\kappa_{\alpha+1}) \le G_\alpha$.

Indeed, suppose β is an ordinal number at most m and κ_α have already been constructed for each $\alpha < \beta$. If $\beta = \alpha + 1$ is a successor ordinal, use Lemma 14.2 to construct a Galois approximation $\kappa_\beta\colon \mathbf{G} \to \mathrm{Gal}(\mathbf{K}_\beta)$ and a rigid epimorphism $\rho_{\beta,\alpha}\colon \mathrm{Gal}(\mathbf{K}_\beta) \to \mathrm{Gal}(\mathbf{K}_\alpha)$ with $\rho_{\beta,\alpha} \circ \kappa_\beta = \kappa_\alpha$ such that K_β is perfect, $K_\alpha \subseteq K_\beta$, $\rho_{\beta,\alpha}\colon \mathrm{Gal}(K_\beta) \to \mathrm{Gal}(K_\alpha)$ is the restriction map, and $\mathrm{Ker}(\kappa_\beta) \le G_\alpha$. If β is a limit ordinal, then $\{\mathrm{Gal}(\mathbf{K}_\alpha), \rho_{\alpha',\alpha} \mid \alpha \leq \alpha' < \beta\}$ is an inverse system of Galois group structures with $K_\alpha \subseteq K_{\alpha'}$, $\rho_{\alpha,\alpha'}: \mathrm{Gal}(K'_\alpha) \to \mathrm{Gal}(K_\alpha)$ are the restriction maps, and $\rho_{\alpha,\alpha'}: \mathrm{Gal}(\mathbf{K}'_\alpha) \to \mathrm{Gal}(\mathbf{K}_\alpha)$ are rigid epimorphisms. Then $\mathrm{Gal}(\mathbf{K}_\beta) = \varprojlim \mathrm{Gal}(\mathbf{K}_\alpha)$ is a group structure with $K_\beta = \bigcup_{\alpha<\beta} K_\alpha$ and with rigid projections $\rho_{\beta,\alpha}: \mathrm{Gal}(\mathbf{K}_\beta) \to \mathrm{Gal}(\mathbf{K}_\alpha)$ (Remark 2.7). Moreover, the inverse limit of the κ_α's gives a Galois approximation $\kappa_\beta: \mathbf{G} \to \mathrm{Gal}(\mathbf{K}_\beta)$ with $\rho_{\beta,\alpha} \circ \kappa_\beta = \kappa_\alpha$ for each $\alpha < \beta$.

Having completed the transfinite induction, we put $\mathbf{L} = \mathbf{K}_m$ and $\lambda = \kappa_m$. Then the underlying field of $\mathbf{L}$ is perfect and $\lambda: \mathbf{G} \to \mathrm{Gal}(\mathbf{L})$ is a Galois approximation lifting κ (Remark 2.7). Moreover, $\mathrm{Ker}(\lambda) \leq \bigcap_{\alpha<m} G_\alpha = 1$. Since $\mathbf{G}$ is proper, λ is an isomorphism (Remark 2.1). □

Remark 14.3. *Cardinality of L.* We may assume that the cardinality of L in Proposition 14.4 is not smaller than any given cardinality m. Indeed, without loss κ is an isomorphism. Hence, if λ lifts κ, then λ is an isomorphism. Put $\lambda_0 = \kappa$. By transfinite induction construct a family of Galois approximations $\lambda_\alpha: \mathbf{G} \to \mathrm{Gal}(\mathbf{L}_\alpha)$ with underlying fields L_α such that λ_β lifts λ_α and $L_\alpha \subset L_\beta$ for all $\alpha \leq \beta \leq m$. Namely, if β is a limit ordinal, put $L_\beta = \bigcup_{\alpha<\beta} L_\beta$ and $L_{\beta,x} = \bigcup_{\alpha<\beta} L_{\alpha,x}$; otherwise use Lemma 14.2 to construct a lifting λ_β of $\lambda_{\beta-1}$. Then $\lambda = \lambda_m$ has the required property. □

15. From Field Structures to Field-Valuation Structures

Having lifted a given Galois approximation $\kappa: \mathbf{G} \to \mathrm{Gal}(\mathbf{K})$ of a proper projective group structure to a Galois isomorphism $\epsilon: \mathbf{G} \to \mathrm{Gal}(\mathbf{E})$, we wish to extend $\mathbf{E}$ to a proper field-valuation structure $\mathbf{L}$ which satisfies the block approximation condition and $\mathrm{res}: \mathrm{Gal}(\mathbf{L}) \to \mathrm{Gal}(\mathbf{K})$ is an isomorphism.

The crucial step of the construction is, starting from a field-valuation structure $\mathbf{K}$ and a data $(V, X_i, L_i, \mathbf{b}_i)_{i\in I_0}$ satisfying (2) below, to extend $\mathbf{K}$ to a field-valuation structure with a point $\mathbf{z} \in V(K')$ blockwise approximating each $\mathbf{b}_i$ infinitely well over K; that is, $\mathbf{z}$ satisfies Condition (3c) below.

Let $\mathbf{K} = (K, X, K_x, v_x)_{x\in X}$ and $\mathbf{K}' = (K', X', K'_x, v'_x)_{x\in X}$ be field-valuation structures. We say **$\mathbf{K}'$ extends $\mathbf{K}$** and write $\mathbf{K} \subseteq \mathbf{K}'$ if $K \subseteq K'$, $K_x \subseteq K'_x$, and $v_x = v'_x|_{K_x}$ for each $x \in X$.

Lemma 15.1. *Let $\mathbf{K} = (K, X, K_x, v_x)_{x\in X}$ and $\bar{\mathbf{K}} = (\bar{K}, X, \bar{K}_x, \bar{v}_x)_{x\in X}$ be proper Henselian field-valuation structures satisfying this:*

(1a) *$\bar{K}$ and K are perfect.*

(1b) *$\bar{\mathbf{K}} \subseteq \mathbf{K}$ and the map $\mathrm{res}_{K_s/\bar{K}_s}: \mathrm{Gal}(\mathbf{K}) \to \mathrm{Gal}(\bar{\mathbf{K}})$ (with the identity map $X \to X$) is an isomorphism.*

(1c) *$\mathrm{Gal}(\bar{\mathbf{K}})$ is projective.*

(1d) *$\bar{v}_x$ is the trivial valuation of $\bar{K}_x$, $x \in X$.*

(1e) *$\bar{K}_x$ is the residue field of (K_x, v_x), $x \in X$.*

Consider a data $(V, X_i, L_i, \mathbf{b}_i)_{i\in I_0}$ satisfying this:

(2a) *$(\mathrm{Gal}(L_i), X_i)_{i\in I_0}$ is a special partition of $\mathrm{Gal}(\mathbf{K})$.*

(2b) *V is a smooth absolutely irreducible affine variety over K.*

(2c) *$\mathbf{b}_i \in V(L_i)$.*

Then $\mathbf{K}$ has a proper field-valuation extension $\mathbf{K}' = (K', X, K'_x, v'_x)_{x\in X}$ with K' perfect satisfying this:

(3a) *(K'_x, v'_x) is a Henselian field with residue field $\bar{K}_x$, $x \in X$.*

(3b) $\mathrm{res}_{\tilde{K}'/\tilde{K}}\colon \mathrm{Gal}(K') \to \mathrm{Gal}(K)$ *together with the identity map* $X \to X$ *form an isomorphism* $\mathrm{Gal}(\mathbf{K}') \to \mathrm{Gal}(\mathbf{K})$.

(3c) *There is a* $\mathbf{z} \in V(K')$ *with* $v'_x(\mathbf{z} - \mathbf{b}_i) > v'_x(c)$ *for all* $i \in I_0$, $x \in X_i$, *and* $c \in K^\times$.

PROOF. First suppose $X = \{x\}$. By Remark 2.1, $K_x = K$. Let

$$(V, X_i, L_i, \mathbf{b}_i)_{i \in I_0}$$

be a data satisfying (2). Then $I_0 = \{i\}$ and $L_i = K$. Hence, $\mathbf{K}' = \mathbf{K}$ and $\mathbf{z} = \mathbf{b}_i$ satisfy (3). We may therefore suppose X has at least two elements.

We construct an extension F of K of large transcendence degree such that $V(F)$ contains a generic point $\mathbf{z}$ of V over K. Then we extend F to a proper field-valuation structure $\mathbf{F} = (F, Y, F_y, w_y)$ with a cover $\pi\colon \mathrm{Gal}(\mathbf{F}) \to \mathrm{Gal}(\mathbf{K})$ such that (F_y, w_y) is a rigid Henselian extension of $(K_{\pi(y)}, v_{\pi(y)})$, $y \in Y$. Since $\mathrm{Gal}(\mathbf{K})$ is projective, $\mathbf{F}$ has a extension $\mathbf{F}' = (K', X', F_y, w_y)_{y \in X'}$ such that $\pi\colon \mathrm{Gal}(\mathbf{K}') \to \mathrm{Gal}(\mathbf{K})$ is an isomorphism. Renaming X' as X gives the desired extension $\mathbf{K}'$ of $\mathbf{K}$. In this construction, the valuations w_y are defined in such a manner that $\mathbf{z}$ blockwise approximates the $\mathbf{b}_i$'s infinitely well over K. The construction has six parts.

PART A. *The field* F. Let $\mathbf{z}$ be a generic point of V over K. Put $E = K(\mathbf{z})_{\mathrm{ins}}$. Since V is absolutely irreducible and K is perfect, E/K is a regular extension. Hence, $\mathrm{res}\colon \mathrm{Gal}(E) \to \mathrm{Gal}(K)$ is an epimorphism. Let $i \in I_0$. By [JaR, p. 456, Cor. A2], there is an L_i-place $\bar{\rho}_i\colon L_i(\mathbf{z}) \to L_i \cup \{\infty\}$ with $\bar{\rho}_i(\mathbf{z}) = \mathbf{b}_i$. By Proposition 7.4, there is a perfect algebraic extension E_i of $L_i(\mathbf{z})$ and an extension of $\bar{\rho}_i$ to a rigid Henselian place $\rho_i\colon E_i \to L_i \cup \{\infty\}$. In particular, $E \subseteq E_i$ and $\rho_i(\mathbf{z}) = \mathbf{b}_i$.

Choose a set T of indeterminates with $\mathrm{card}(T) \ge \mathrm{card}(E)$. Put $F = E(T)_{\mathrm{ins}}$. Then F is a regular extension of E, hence of K. Therefore, $\mathrm{res}\colon \mathrm{Gal}(F) \to \mathrm{Gal}(K)$ is an epimorphism. In addition, $\mathbf{z} \in V(F)$.

PART B. *The field structure* $(F, Y, F_y)_{y \in Y}$. Lemma 13.3 gives for each $i \in I_0$ a valuation w'_i of F with residue field E_i and a Henselian closure (F_i, w'_i) of (F, w'_i) such that the corresponding place $\varphi_i\colon F_i \to E_i \cup \{\infty\}$ is rigid.

Put $\psi_i = \rho_i \circ \varphi_i$. Then $\psi_i\colon F_i \to L_i \cup \{\infty\}$ is a rigid L_i-place (Lemma 13.2(c)). In particular, $\mathrm{res}\colon \mathrm{Gal}(F_i) \to \mathrm{Gal}(L_i)$ is an isomorphism. Moreover, ψ_i extends to a $\tilde{K}$-place $\psi_i\colon \tilde{F} \to \tilde{K} \cup \{\infty\}$ with $\psi_i(F') = (F' \cap \tilde{K}) \cup \{\infty\}$ for each algebraic extension F' of F_i. Denote the corresponding valuation by w'_i. Thus, if F' is not algebraically closed, then the residue field of F' with respect to w'_i is not algebraically closed. By Lemma 13.4,

$$F_i F_i^\kappa = \tilde{F} \tag{4}$$

for each $\kappa \in \mathrm{Gal}(F) \smallsetminus \mathrm{Gal}(F_i)$.

By Lemma 5.1, $\mathrm{Gal}(F)$ extends to a proper group structure

$$\mathrm{Gal}(\mathbf{F}) = (\mathrm{Gal}(F), Y, \mathrm{Gal}(F_y))_{y \in Y} \tag{5}$$

and $\mathrm{res}\colon \mathrm{Gal}(F) \to \mathrm{Gal}(K)$ extends to a cover $\pi\colon \mathrm{Gal}(\mathbf{F}) \to \mathrm{Gal}(\mathbf{K})$ of group structures.

PART C. *The field-valuation structure* $\mathbf{F} = (F, Y, F_y, w_y)_{y \in Y}$. In addition to the cover π mentioned in Part B, Lemma 5.1 gives for each $i \in I_0$ a subspace Y_i of Y such that $\pi(Y_i) = X_i$, $F_i \le F_y$ for each $y \in Y_i$, and $Y = \bigcup_{i \in I_0} Y_i^{\mathrm{Gal}(E)}$.

Consider $i \in I_0$ and $y \in Y_i$. Let $x = \pi(y)$. Then $F_i \le F_y$, $L_i \le K_x$, and res: $\mathrm{Gal}(F_y) \to \mathrm{Gal}(K_x)$ is an isomorphism (because π is a cover). Thus, $F_y = F_i K_x$ and $K_x = F_y \cap \tilde{K}$. Since $\psi_i: F_i \to L_i \cup \{\infty\}$ is a rigid L_i-place (Part B), $\psi_i(F_y) = K_x \cup \{\infty\}$ (Lemma 13.2(b)). Since $\mathbf{K}$ is proper and X has at least two elements, $K_x \ne \tilde{K}$ (Remark 2.1), so $F_y \ne \tilde{F}$.

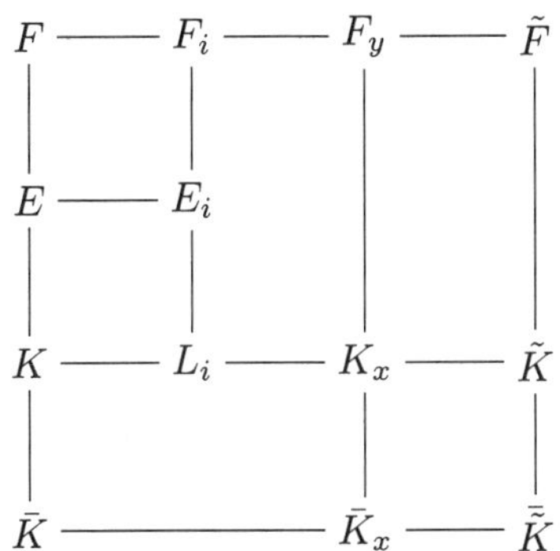

By assumption, (K_x, v_x) is Henselian. Hence, v_x uniquely extends to a valuation v_x of $\tilde{K}$. Let $w_y = v_x \cdot w_i'$ be the unique valuation of $\tilde{F}$ such that $w_y(u) = v_x(\psi_i(u))$ for each $u \in \tilde{F}$ with $\psi_i(u) \in \tilde{K}$. Then $O_{w_y} = \{u \in \tilde{F} \mid \psi_i(u) \in \tilde{K}$ and $v_x(\psi_i(u)) \ge 0\}$, Thus, if $u \in \tilde{F}$ satisfies $\psi_i(u) = \infty$, then $w_y(u) < 0$. If $u \in \tilde{K}$, then $\psi_i(u) = u$, so $w_y(u) = v_x(u)$. Hence, w_y extends v_x (See also the beginning of Section 13.) Since (K_x, v_x) and (F_y, w_i') are Henselian, (F_y, w_y) is Henselian [Jar, Prop. 13.1]. In addition, $\bar{K}_x$ is the residue field of F_y at w_y.

We would like to define (F_y, w_y) for all $y \in Y$. So, we consider $\sigma \in \mathrm{Gal}(F)$ and suppose, in addition to the assumption made above, that $y^\sigma \in Y_j$ for some $j \in I_0$. We prove that $w_{y^\sigma} = w_y^\sigma$.

Indeed, $\pi(y) \in X_i$ and $\pi(y)^{\pi(\sigma)} \in X_j$. Hence, $X_i^{\pi(\sigma)} \cap X_j \ne \emptyset$. By (2a) and by (2g) of Section 3, $i = j$ and $\pi(\sigma) \in \mathrm{Gal}(L_i)$. Hence, there are $\zeta \in \mathrm{Gal}(F_i)$ and $\kappa \in \mathrm{Gal}(F\tilde{K})$ with $\sigma = \kappa\zeta$. Since $y^\sigma \in Y_i$, we have $F_i \subseteq F_{y^\sigma} = F_y^\sigma$. Therefore, $F_i F_i^{\kappa^{-1}} = F_i F_i^{\zeta\sigma^{-1}} = F_i F_i^{\sigma^{-1}} \subseteq F_y \subset \tilde{F}$. By (4), $\kappa = 1$, so $\sigma \in \mathrm{Gal}(F_i)$. Now consider $u \in F_y^\sigma$ with $\psi_i(u) \in K_x$. Since ψ_i is rigid, $\psi_i(u^{\sigma^{-1}}) = \psi_i(u)^{\pi(\sigma)^{-1}}$ (Proposition 7.4(a)). Therefore, $w_y^\sigma(u) = w_y(u^{\sigma^{-1}}) = v_x(\psi_i(u^{\sigma^{-1}})) = v_x(\psi_i(u)^{\pi(\sigma)^{-1}}) = v_x^{\pi(\sigma)}(\psi_i(u)) = v_{x^{\pi(\sigma)}}(\psi_i(u)) = v_{\pi(y^\sigma)}(\psi_i(u)) = w_{y^\sigma}(u)$. It follows, $w_y^\sigma = w_{y^\sigma}$ on F_{y^σ}, and therefore also on $\tilde{F}$, as claimed.

For an arbitrary $y' \in Y$ there are $\tau \in \mathrm{Gal}(F)$, $i \in I_0$, and $y \in Y_i$ with $y' = y^\tau$. Since $(F, Y, F_y)_{y \in Y}$ is a field structure, $F_{y'} = F_y^\tau$. Define $w_{y'}$ to be w_y^τ. By the preceding paragraph, this is a good definition. Thus, with $x' = \pi(y')$, the valued field $(F_{y'}, w_{y'})$ is a rigid Henselian extension of $(K_{x'}, v_{x'})$. Moreover, $w_{(y')^\sigma} = w_{y'}^\sigma$ for all $\sigma \in \mathrm{Gal}(F)$.

PART D. *Continuity of the map* $\nu_F: Y_F \to \mathrm{Val}(\tilde{F})$. For each $x \in X$ let $\nu_K(x) = v_x$. Since $\mathbf{K}$ is a Henselian field-valuation structure, the map $\nu_K: X \to \mathrm{Val}(\tilde{K})$ is continuous (Lemma 11.2). Similarly, for each $y \in Y$ let $\nu_F(y) = w_y$. By Lemma 11.2, it suffices to prove that the map $\nu_F: Y \to \mathrm{Val}(\tilde{F})$ is continuous.

We start by proving that for each $i \in I_0$, the restriction of ν_F to Y_i is continuous. Let $y \in Y_i$ and let $u \in \tilde{F}$ such that $w_y(u) > 0$. By Part C, $\psi_i(u) \ne \infty$ and

$v_{\pi(y)}(\psi_i(u))) = w_y(y) > 0$. If $y' \in Y$ is sufficiently close to y, then $\pi(y')$ is sufficiently close to $\pi(y)$, and hence $v_{\pi(y')}(\psi_i(u))) > 0$ (because ν_K is continuous). Thus $w_{y'}(u) > 0$. Similarly, if $w_y(u) \geq 0$ and y' is sufficiently close to y, then $w_{y'}(u) \geq 0$.

It follows that the map $\nu_i\colon Y_i \times \mathrm{Gal}(F) \to \mathrm{Val}(\tilde{F})$ given by $\nu_i(y,\tau) = w_y^\tau$ is continuous. Indeed, let $a \in \tilde{F}$ and suppose $w_y^\tau(a) > 0$. Then $w_y(a^{\tau^{-1}}) > 0$. If $y' \in Y_i$ is sufficiently close to y and $\tau' \in \mathrm{Gal}(F)$ is sufficiently closed to τ, then, by the preceding paragraph, $w_{y'}^{\tau'}(a) = w_{y'}(a^{(\tau')^{-1}}) = w_{y'}(a^{\tau^{-1}}) > 0$. Similar statement holds for $\geq$ replacing $>$.

Let $\tilde{\nu}_i$ be the restriction of ν_F to $Y_i^{\mathrm{Gal}(F)}$. Let $\mu\colon Y_i \times \mathrm{Gal}(F) \to Y_i^{\mathrm{Gal}(F)}$ be the map defined by $\mu(y,\tau) = y^\tau$. By Part C, $\nu_i = \tilde{\nu}_i \circ \mu$. Also, μ a continuous map between profinite spaces, hence closed. By the preceding paragraph, for each closed subset C of $\mathrm{Val}(F)$, the set $\nu_i^{-1}(C)$ is closed in $Y_i \times \mathrm{Gal}(F)$. Therefore, $\tilde{\nu}_i^{-1}(C) = \mu(\nu_i^{-1}(C))$ is a closed subset of $Y_i^{\mathrm{Gal}(F)}$. Consequently, $\tilde{\nu}_i$ is continuous.

Since $Y = \bigcup_{i\in I_0} Y_i^{\mathrm{Gal}(F)}$, the preceding paragraph implies $\nu_F\colon Y \to \mathrm{Val}(\tilde{F})$ is continuous, as claimed.

PART E. *The proper group structure* $\mathbf{G}'$. By (1b) and (1c), $\mathrm{Gal}(\mathbf{K})$ is projective. By Part B, $\pi\colon \mathrm{Gal}(\mathbf{F}) \to \mathrm{Gal}(\mathbf{K})$ is a cover of group structures. Hence, by Corollary 4.3, $\mathrm{Gal}(\mathbf{F})$ has a proper sub-group-structure

$$\mathbf{G}' = (\mathrm{Gal}(K'), X', \mathrm{Gal}(F_{x'}))_{x'\in X'} ,$$

where K' is an algebraic extension of F and $X' \subseteq Y$ such that $\pi\colon \mathbf{G}' \to \mathrm{Gal}(\mathbf{K})$ is an isomorphism. In particular, $\mathrm{res}\colon \mathrm{Gal}(K') \to \mathrm{Gal}(K)$ is an isomorphism and $\pi\colon X' \to X$ is a homeomorphism. Then $\mathbf{F}' = (K', X', F_{x'}, w_{x'})_{x'\in X'}$ is a field-valuation structure.

PART F. *The proper field-valuation structure* $\mathbf{K}'$. For each $x \in X$ let x' be the unique element of X' with $\pi(x') = x$. Put $K'_x = F_{x'}$ and $v'_x = w_{x'}$. Then $\mathbf{K}' = (K', X, K'_x, v'_x)_{x\in X}$ is a proper field structure isomorphic to $\mathbf{F}'$. In addition, $\mathbf{K}'$ extends $\mathbf{K}$ and satisfies Conditions (3a) and (3b).

We still have to prove Condition (3c) (block approximation). Let $\mathbf{z} = (z_1, \dots, z_n)$ and $\mathbf{b}_i = (b_{i1}, \dots, b_{in})$, $i \in I_0$. Then $\psi_i(\mathbf{z}) = \rho_i(\mathbf{z}) = \mathbf{b}_i$, $\psi(\mathbf{b}_i) = \mathbf{b}_i$, and $\psi(c) = c$ for all $c \in \tilde{K}$. Let $y \in Y_i$ and put $x = \pi(y)$. Then, for all $c \in K^\times$ and $1 \leq j \leq n$ we have

$$w_y\Big(\frac{z_j - b_{ij}}{c}\Big) = v_x\Big(\frac{\psi_i(z_j) - b_{ij}}{c}\Big) = v_x\Big(\frac{0}{c}\Big) > 0.$$

Therefore, $w_y(\mathbf{z} - \mathbf{b}_i) > w_y(c)$.

Finally, consider $x \in X_i$. Choose $x' \in X'$ and $y \in Y_i$ with $\pi(x') = x = \pi(y)$. Then there is a $\kappa \in \mathrm{Gal}(F\tilde{K})$ with $x' = y^\kappa$, hence $\mathbf{z}^{\kappa^{-1}} = \mathbf{z}$ and $\mathbf{b}^{\kappa^{-1}} = \mathbf{b}$. By the preceding paragraph, $v'_x(\mathbf{z} - \mathbf{b}_i) = w_{x'}(\mathbf{z} - \mathbf{b}_i) = w_y(\mathbf{z}^{\kappa^{-1}} - \mathbf{b}^{\kappa^{-1}}) = w_y(\mathbf{z} - \mathbf{b}_i) > w_y(c) = v_x(c) = v'_x(c)$ for all $c \in K^\times$. This concludes the proof of the Lemma. □

We apply Lemma 15.1 in each step of a transfinite induction. In the rest of this section we write $\mathrm{res}\colon \mathrm{Gal}(\mathbf{L}) \to \mathrm{Gal}(\mathbf{K})$ for proper field structures $\mathbf{K} \subseteq \mathbf{L}$ to denote the unique morphism that extend the homomorphism $\mathrm{res}\colon \mathrm{Gal}(L) \to \mathrm{Gal}(K)$ (Remark 2.1).

LEMMA 15.2. *Let* $\mathbf{K} = (K, X, K_x, v_x)_{x\in X}$ *and* $\bar{\mathbf{K}} = (\bar{K}, X, \bar{K}_x, \bar{v}_x)$ *be proper Henselian field-valuation structures satisfying (1). Then* $\mathbf{K}$ *has a proper field-valuation extension* $\mathbf{L} = (L, X, L_x, w_x)_{x\in X}$ *with* L *perfect satisfying this:*

(6a) (L_x, w_x) *is Henselian with residue field* $\bar{K}_x$, $x \in X$.

(6b) $\mathrm{res}\colon \mathrm{Gal}(\mathbf{L}) \to \mathrm{Gal}(\mathbf{K})$ *is an isomorphism.*

(6c) $\mathbf{L}$ *satisfies the block approximation condition.*

PROOF. Well-order all data satisfying (2) in a transfinite sequence

$$(V_\alpha, X_{\alpha,i}, K_{\alpha,i}, \mathbf{b}_{\alpha,i})_{i\in I_\alpha}, \quad \alpha < m.$$

Use transfinite induction and Lemma 15.1 to construct for each ordinal number $\alpha \le m$ a proper field-valuation structure $\mathbf{K}_\alpha = (K_\alpha, X, K_{\alpha,x}, v_{\alpha,x})_{x\in X}$ with K_α perfect satisfying these conditions:

(7a) $(K_{\alpha,x}, v_{\alpha,x})$ is a Henselian field with residue field $\bar{K}_x$, $x \in X$.

(7b) $\mathbf{K}_\alpha \subseteq \mathbf{K}_\beta$ and $\mathrm{res}\colon \mathrm{Gal}(\mathbf{K}_\beta) \to \mathrm{Gal}(\mathbf{K}_\alpha)$ is an isomorphism for all $\alpha < \beta \le m$.

(7c) $\mathbf{K}_\beta = \bigcup_{\alpha<\beta} \mathbf{K}_\alpha$ for each limit ordinal $\beta \le m$.

(7d) For each ordinal number $\alpha < m$ there is a point $\mathbf{z} \in V_\alpha(K_{\alpha+1})$ with $v_{\alpha+1,x}(\mathbf{z} - \mathbf{b}_{\alpha,i}) > v_{\alpha+1,x}(c)$ for all $i \in I_\alpha$, $x \in X_{\alpha,i}$, and $c \in K_\alpha^\times$.

Rewrite $\mathbf{K}_m$ as $\mathbf{L}_1 = (L_1, X, L_{1,x}, w_{1,x})_{x\in X}$. Then:

(8a) $(L_1, v_{1,x})$ is a Henselian field with residue field $\bar{K}_x$, $x \in X$.

(8b) $\mathbf{K} \subseteq \mathbf{L}_1$ and $\mathrm{res}\colon \mathrm{Gal}(\mathbf{L}_1) \to \mathrm{Gal}(\mathbf{K})$ is an isomorphism.

(8c) Each approximation problem $(V, X_i, K_i, \mathbf{b}_i)_{i\in I_0}$ for $\mathbf{K}$ has a solution $\mathbf{z} \in V(L_1)$.

Finally use usual induction to construct an ascending sequence of proper field-valuation structures $\mathbf{L}_j$, $j = 1, 2, 3, \ldots$, such that $\mathbf{L}_{j+1}$ relates to $\mathbf{L}_j$ in the same way that $\mathbf{L}_1$ relates to $\mathbf{K}$, $j = 1, 2, 3, \ldots$. The structure $\mathbf{L} = \bigcup_{j=1}^\infty \mathbf{L}_j$ satisfies (6).□

PROPOSITION 15.3. *Let* $\mathbf{K} = (K, X, K_x)_{x\in X}$ *be a proper field structure with* $\mathrm{Gal}(\mathbf{K})$ *projective. Then there is a proper field-valuation structure* $\mathbf{L} = (L, X, L_x, w_x)_{x\in X}$ *with* L *perfect having these properties:*

(9a) $(L, X, L_x)_{x\in X}$ *extends* $\mathbf{K}$.

(9b) (L_x, w_x) *is Henselian with residue field* $(K_x)_{\mathrm{ins}}$, $x \in X$.

(9c) $\mathrm{res}\colon \mathrm{Gal}(\mathbf{L}) \to \mathrm{Gal}(\mathbf{K})$ *is an isomorphism.*

(9d) $\mathbf{L}$ *satisfies the block approximation condition.*

PROOF. Replace $\mathbf{K}$ by $\mathbf{K}_{\mathrm{ins}}$, if necessary, to assume K is perfect. Identify $\mathbf{K}$ with $(K, X, K_x, v_x)_{x\in X}$, where v_x the trivial valuation on K_x for each $x \in X$. Put $\bar{K}_x = K_x$, $\bar{v}_x = v_x$, $\bar{K} = K$, and $\bar{\mathbf{K}} = \mathbf{K}$. Then $(\bar{\mathbf{K}}, \mathbf{K})$ satisfies (1). Lemma 15.2 gives $\mathbf{L}$ satisfying (9). □

We are finally ready to prove Part (b) of the Main Theorem:

THEOREM 15.4. *Let* $\mathbf{K}$ *be a field structure,* $\mathbf{G}$ *a projective group structure, and* $\kappa\colon \mathbf{G} \to \mathrm{Gal}(\mathbf{K})$ *a Galois approximation. Then there exists a proper Henselian field-valuation structure* $\mathbf{L} = (L, X, L_x, w_x)_{x\in X}$ *and an isomorphism* $\psi\colon \mathbf{G} \to \mathrm{Gal}(\mathbf{L})$ *with* L *perfect having these properties:*

(10a) $\mathbf{K} \subseteq \mathbf{L}$ *and* $\mathrm{res} \circ \psi = \kappa$.

(10b) w_x *is trivial on* K, $x \in X$.

(10c) $\mathbf{L}$ *satisfies the block approximation condition.*

PROOF. Replace K by K_x and K_{ins} by $(K_x)_{\mathrm{ins}}$, if necessary, to assume K is perfect. Proposition 14.4 gives a proper field structure $\mathbf{K}'$ which extends $\mathbf{K}$ and an isomorphism $\kappa'\colon \mathbf{G} \to \mathrm{Gal}(\mathbf{K}')$ with $\mathrm{res}_{\tilde{K}'/\tilde{K}} \circ \kappa' = \kappa$. Proposition 15.3 extends $\mathbf{K}'$ to a proper Henselian field-valuation structure $\mathbf{L} = (L, X, L_x, w_x)_{x \in X}$ that satisfies the block approximation theorem such that L is perfect and $\mathrm{res}_{\tilde{L}/\tilde{K}'}\colon \mathrm{Gal}(\mathbf{L}) \to \mathrm{Gal}(\mathbf{K}')$ is an isomorphism. Thus, there is an isomorphism $\psi\colon \mathbf{G} \to \mathrm{Gal}(\mathbf{L})$ with $\mathrm{res}_{L_s/K'_s} \circ \psi = \varphi'$. This establishes (10a), (10b), and (10c). □

An easy consequence of Theorem 15.4 is the realization of profinite products of finitely many absolute Galois groups as an absolute Galois group. Of course, one may get away with a much reduced machinery than the one we have developed here. See [Ers], [Koe], or [HJK].

THEOREM 15.5. *For each i in a set I_0 let K_i be a field which is not separably closed. Then there is a proper Henselian field-valuation structure $\mathbf{L} = (L, X, L_x, w_x)_{x \in X}$ with* $\mathrm{char}(L) = 0$ *satisfying the block approximation condition and* $G(L) \cong \mathop{\ast\!\!\!\!\prod}_{i \in I_0} \mathrm{Gal}(K_i)$.

PROOF. Choose a set T of cardinality at least the transcendence degree of K_i over its prime field for all $i \in I_0$. By Proposition 7.5, $\mathbb{Q}(T)$ has an algebraic extension K_i' with $\mathrm{Gal}(K_i) \cong \mathrm{Gal}(K_i')$. Let $K = \bigcap_{i \in I_0} K_i'$. Then replace K_i by K_i', if necessary, to assume all K_i are algebraic extension of K and $\mathrm{Gal}(K) = \langle \mathrm{Gal}(K_i) \mid i \in I_0 \rangle$.

For each $i \in I_0$ let G_i be an isomorphic copy of $\mathrm{Gal}(K_i)$ and $\kappa_i\colon G_i \to \mathrm{Gal}(K_i)$ an isomorphism. Example 4.7(c) constructs a proper projective group structure $\mathbf{G} = (G, X, G_x)_{x \in X}$ with $G = \mathop{\ast\!\!\!\!\prod}_{i \in I_0} \mathrm{Gal}(K_i)$. Let $\kappa\colon G \to \mathrm{Gal}(K)$ be the epimorphism whose restriction to G_i is κ_i. By Example 2.5, $\mathbf{G}/\mathrm{Ker}(\kappa)$ is a group structure and the quotient map $\mathbf{G} \to \mathbf{G}/\mathrm{Ker}(\kappa)$ is a cover. Thus, there is a field structure $\mathbf{K} = (K, Y, K_y)$ and κ extend to a cover $\kappa\colon \mathbf{G} \to \mathrm{Gal}(\mathbf{K})$. Theorem 15.4 gives the desired field-valuation structure $\mathbf{L}$. □

Bibliography

[ArK] R. F. Arens and I. Kaplansky, *Topological representations of algebras*, Transactions of AMS **63** (1948), 457–481.

[Art] E. Artin, *Algebraic Numbers and Algebraic Functions*, Gordon and Breach, New York, 1967.

[Bre] G. E. Bredon, *Introduction to Compact Transformation Groups*, Pure and applied Mathematics **46**, Academic Press, New Your, 1972.

[Bou] N. Bourbaki, *Commutative Algebra, Chapters 1–7*, Springer, Berlin, 1989.

[Efr] I. Efrat, *A Galois-theoretic characterization of p-adically closed fields*, Israel Journal of Mathematics **91** (1995), 273-284.

[End] O. Endler, *Valuation Theory*, Springer, Berlin, 1972.

[FHV] M. D. Fried, D. Haran, H. Völklein, *Real hilbertianity and the field of totally real numbers*, Contemporary Mathematics **174** (1994), 1–34.

[FrJ] M. D. Fried and M. Jarden, *Field Arithmetic, Second Edition, revised and enlarged by Moshe Jarden*, Ergebnisse der Mathematik (3) **11**, Springer, Heidelberg, 2005.

[Gil] R. Gill-more, *On free products of profinite groups*, MSc Thesis, Tel Aviv, 1995.

[GPR] B. Green, F. Pop, and P. Roquette, *On Rumely's local-global principle*, Jahresbericht der Deutschen Mathematiker-Vereinigung **97** (1995), 43–74.

[Har] D. Haran, *On closed subgroups of free products of profinite groups*, Proceedings of the London Mathematical Society **55** (1987), 266–289.

[HaJ1] D. Haran and M. Jarden, *The absolute Galois group of a pseudo real closed field*, Annali della Scuola Normale Superiore — Pisa, Serie IV, **12** (1985), 449–489.

[HaJ2] D. Haran and M. Jarden, *The absolute Galois group of a pseudo p-adically closed field*, Journal für die reine und angewandte Mathematik **383** (1988), 147–206.

[HaJ3] D. Haran and M. Jarden, *Relatively projective groups as absolute Galois groups*, in "Progress in Galois Theory", edts. H. Voelklein and T. Shaska, pp. 87-1000, Springer Science, 2005.

[HJK] D. Haran, M. Jarden, and J. Koenigsmann, *Free products of absolute Galois groups*, manuscript, Tel Aviv, 2000, http://www.tau.ac.il/~jarden.

[Hrt] R. Hartshorne, *Algebraic Geometry*, Graduate Texts in Mathematics **52**, Springer, New York, 1977.

[HeR] W. Herfort and L. Ribes, *Torsion elements and centralizers in free products of profinite groups,* Journal für die reine und angewandte Mathematik **358** (1985), 155-161.

[HRo] E. Hewitt and K. A. Ross, *Abstract Harmonic Analysis I,* Die Grundlehren der Mathematischen Wissenschaften **115**, Springer-Verlag, Berlin, 1963.

[Hoe] M. Hochster, *Prime ideal structure in commutative rings,* Transactions of American Mathematical Society, **142** (1969), 43-60.

[Jar] M. Jarden, *Intersection of local algebraic extensions of a Hilbertian field (A. Barlotti et al., eds),* NATO ASI Series C **333** 343–405, Kluwer, Dordrecht, 1991.

[JaR] M. Jarden and Peter Roquette, *The Nullstellensatz over p-adically closed fields,* Journal of the Mathematical Society of Japan **32** (1980), 425–460.

[Koe] J. Koenigsmann, *Relatively projective groups as absolute Galois groups,* Israel Journal of Mathematics **127** (2002), 93–129.

[KPR] F.-V. Kuhlmann, M. Pank, and P. Roquette, *Immediate and purely wild extensions of valued fields,* manuscripta mathematicae **55** (1986), 39–67.

[Mum] D. Mumford, *The Red Book of Varieties and Schemes,* Lecture Notes in Mathematics **1358**, Springer, Berlin, 1988.

[Pop] F. Pop, *Classically projective groups,* preprint, Heidelberg, 1990.

[Pre] A. Prestel, *On the axiomatization of PRC-fields,* in Methods of Mathematical Logic, Lecture Notes in Mathematics **1130** (1985), 351–359.

[Ray] M. Raynaud, *Anneaux Locaux Henséliens,* Lecture Notes in Mathematics **169**, Springer, Berlin.

[Rib] L. Ribes, *Introduction to Profinite Groups and Galois Cohomology,* Queen's papers in Pure and Applied Mathematics **24**, Queen's University, Kingston, 1970.

Editorial Information

To be published in the *Memoirs*, a paper must be correct, new, nontrivial, and significant. Further, it must be well written and of interest to a substantial number of mathematicians. Piecemeal results, such as an inconclusive step toward an unproved major theorem or a minor variation on a known result, are in general not acceptable for publication.

Papers appearing in *Memoirs* are generally at least 80 and not more than 200 published pages in length. Papers less than 80 or more than 200 published pages require the approval of the Managing Editor of the Transactions/Memoirs Editorial Board.

As of May 31, 2007, the backlog for this journal was approximately 15 volumes. This estimate is the result of dividing the number of manuscripts for this journal in the Providence office that have not yet gone to the printer on the above date by the average number of monographs per volume over the previous twelve months, reduced by the number of volumes published in four months (the time necessary for preparing a volume for the printer). (There are 6 volumes per year, each usually containing at least 4 numbers.)

A Consent to Publish and Copyright Agreement is required before a paper will be published in the *Memoirs*. After a paper is accepted for publication, the Providence office will send a Consent to Publish and Copyright Agreement to all authors of the paper. By submitting a paper to the *Memoirs*, authors certify that the results have not been submitted to nor are they under consideration for publication by another journal, conference proceedings, or similar publication.

Information for Authors

Memoirs are printed from camera copy fully prepared by the author. This means that the finished book will look exactly like the copy submitted.

Initial submission. The AMS uses Centralized Manuscript Processing for initial submissions. Authors should submit a PDF file using the Initial Manuscript Submission form found at `www.ams.org/cgi-bin/peertrack/submission.pl`, or send one copy of the manuscript to the following address: Centralized Manuscript Processing, MEMOIRS OF THE AMS, 201 Charles Street, Providence, RI 02904-2294 USA. If a paper copy is being forwarded to the AMS, indicate that it is for it Memoirs and include the name of the corresponding author, contact information such as email address or mailing address, and the name of an appropriate Editor to review the paper (see the list of Editors below).

The paper must contain a *descriptive title* and an *abstract* that summarizes the article in language suitable for workers in the general field (algebra, analysis, etc.). The *descriptive title* should be short, but informative; useless or vague phrases such as "some remarks about" or "concerning" should be avoided. The *abstract* should be at least one complete sentence, and at most 300 words. Included with the footnotes to the paper should be the 2000 *Mathematics Subject Classification* representing the primary and secondary subjects of the article. The classifications are accessible from `www.ams.org/msc/`. The list of classifications is also available in print starting with the 1999 annual index of *Mathematical Reviews*. The Mathematics Subject Classification footnote may be followed by a list of *key words and phrases* describing the subject matter of the article and taken from it. Journal abbreviations used in bibliographies are listed in the latest *Mathematical Reviews* annual index. The series abbreviations are also accessible from `www.ams.org/publications/`. To help in preparing and verifying references, the AMS offers MR Lookup, a Reference Tool for Linking, at `www.ams.org/mrlookup/`.

Electronically prepared manuscripts. The AMS encourages electronically prepared manuscripts, with a strong preference for $\mathcal{A}_{\mathcal{M}}\mathcal{S}$-LaTeX. To this end, the Society has prepared $\mathcal{A}_{\mathcal{M}}\mathcal{S}$-LaTeX author packages for each AMS publication. Author packages include instructions for preparing electronic manuscripts, samples, and a style file that generates the particular design specifications of that publication series. Though $\mathcal{AMS}$-LaTeX is the highly preferred format of TeX, author packages are also available in $\mathcal{AMS}$-TeX.

Authors may retrieve an author package from the AMS website starting from `www.ams.org/tex/` or via FTP to `ftp.ams.org` (login as `anonymous`, enter username as password, and type `cd pub/author-info`). The *AMS Author Handbook* and the *Instruction Manual* are available in PDF format following the author packages link from `www.ams.org/tex/`. The author package can also be obtained free of charge by sending email to `tech-support@ams.org` (Internet) or from the Publication Division, American Mathematical Society, 201 Charles St., Providence, RI 02904-2294, USA. When requesting an author package, please specify $\mathcal{AMS}$-LaTeX or $\mathcal{AMS}$-TeX and the publication in which your paper will appear. Please be sure to include your complete mailing address.

After acceptance. The final version of the electronic file should be sent to the Providence office (this includes any TeX source file, any graphics files, and the DVI or PostScript file) immediately after the paper has been accepted for publication.

Before sending the source file, be sure you have proofread your paper carefully. The files you send must be the EXACT files used to generate the proof copy that was accepted for publication. For all publications, authors are required to send a printed copy of their paper, which exactly matches the copy approved for publication, along with any graphics that will appear in the paper.

Accepted electronically prepared files can be submitted via the web at `www.ams.org/submit-book-journal/`, sent via FTP, or sent on CD-Rom or diskette to the Electronic Prepress Department, American Mathematical Society, 201 Charles Street, Providence, RI 02904-2294 USA. TeX source files, DVI files, and PostScript files can be transferred over the Internet by FTP to the Internet node `ftp.ams.org` (130.44.1.100). When sending a manuscript electronically via CD-Rom or diskette, please be sure to include a message identifying the paper as a Memoir.

Electronically prepared manuscripts can also be sent via email to `pub-submit@ams.org` (Internet). In order to send files via email, they must be encoded properly. (DVI files are binary and PostScript files tend to be very large.)

Electronic graphics. Comprehensive instructions on preparing graphics are available at `www.ams.org/jourhtml/`. A few of the major requirements are given here.

Submit files for graphics as EPS (Encapsulated PostScript) files. This includes graphics originated via a graphics application as well as scanned photographs or other computer-generated images. If this is not possible, TIFF files are acceptable as long as they can be opened in Adobe Photoshop or Illustrator. No matter what method was used to produce the graphic, it is necessary to provide a paper copy to the AMS.

Authors using graphics packages for the creation of electronic art should also avoid the use of any lines thinner than 0.5 points in width. Many graphics packages allow the user to specify a "hairline" for a very thin line. Hairlines often look acceptable when proofed on a typical laser printer. However, when produced on a high-resolution laser imagesetter, hairlines become nearly invisible and will be lost entirely in the final printing process.

Screens should be set to values between 15% and 85%. Screens which fall outside of this range are too light or too dark to print correctly. Variations of screens within a graphic should be no less than 10%.

Inquiries. Any inquiries concerning a paper that has been accepted for publication should be sent to `memo-query@ams.org` or directly to the Electronic Prepress Department, American Mathematical Society, 201 Charles St., Providence, RI 02904-2294 USA.

Editors

Titles in This Series

887 **Charlotte Wahl,** Noncommutative Maslov index and eta-forms, 2007

886 **Robert M. Guralnick and John Shareshian,** Symmetric and alternating groups as monodromy groups of Riemann surfaces I: Generic covers and covers with many branch points, 2007

885 **Jae Choon Cha,** The structure of the rational concordance group of knots, 2007

884 **Dan Haran, Moshe Jarden, and Florian Pop,** Projective group structures as absolute Galois structures with block approximation, 2007

883 **Apostolos Beligiannis and Idun Reiten,** Homological and homotopical aspects of torsion theories, 2007

882 **Lars Inge Hedberg and Yuri Netrusov,** An axiomatic approach to function spaces, spec tral synthesis and Luzin approximation, 2007

881 **Tao Mei,** Operator valued Hardy spaces, 2007

880 **Bruce C. Berndt, Geumlan Choi, Youn-Seo Choi, Heekyoung Hahn, Boon Pin Yeap, Ae Ja Yee, Hamza Yesilyurt, and Jinhee Yi,** Ramanujan's forty identities for Rogers-Ramanujan functions, 2007

879 **O. García-Prada, P. B. Gothen, and V. Muñoz,** Betti numbers of the moduli space of rank 3 parabolic Higgs bundles, 2007

878 **Alessandra Celletti and Luigi Chierchia,** KAM stability and celestial mechanics, 2007

877 **María J. Carro, José A. Raposo, and Javier Soria,** Recent developments in the theory of Lorentz spaces and weighted inequalities, 2007

876 **Gabriel Debs and Jean Saint Raymond,** Borel liftings of Borel sets: Some decidable and undecidable statements, 2007

875 **C. Krattenthaler and T. Rivoal,** Hypergéométrie et fonction zêta de Riemann, 2007

874 **Sonia Natale,** Semisolvability of semisimple Hopf algebras of low dimension, 2007

873 **A. J. Duncan,** Exponential genus problems in one-relator products of groups, 2007

872 **Anthony V. Geramita, Tadahito Harima, Juan C. Migliore, and Yong Su Shin,** The Hilbert function of a level algebra, 2007

871 **Pascal Auscher,** On necessary and sufficient conditions for L^p-estimates of Riesz transforms associated to elliptic operators on $\mathbb{R}^n$ and related estimates, 2007

870 **Takuro Mochizuki,** Asymptotic behaviour of tame harmonic bundles and an application to pure twistor D-modules, Part 2, 2007

869 **Takuro Mochizuki,** Asymptotic behaviour of tame harmonic bundles and an application to pure twistor D-modules, Part 1, 2007

868 **Gelu Popescu,** Entropy and multivariable interpolation, 2006

867 **Vilmos Totik,** Metric properties of harmonic measures, 2006

866 **William Craig,** Semigroups underlying first-order logic, 2006

865 **Nathanial P. Brown,** Invariant means and finite representation theory of $C*$-algebras, 2006

864 **John M. Lee,** Fredholm operators and Einstein metrics on conformally compact manifolds, 2006

863 **M. Lübke and A. Teleman,** The Universal Kobayashi-Hitchin correspondence on Hermitian manifolds, 2006

862 **Alberto Canonaco,** The Beilinson complex and canonical rings of irregular surfaces, 2006

861 **Leon A. Takhtajan and Lee-Peng Teo,** Weil-Petersson metric on the universal Teichmüller space, 2006

860 **Thomas M. Fiore,** Pseudo limits, biadjoints and pseudo algebras: Categorical foundations of conformal field theory, 2006

859 **N. Arcozzi, R. Rochberg, and E. Sawyer,** Carleson measures and interpolating sequences for Besov spaces on complex balls, 2006

858 **Enrico Valdinoci, Berardino Sciunzi, and Vasile Ovidiu Savin,** Flat level set regularity of p-Laplace phase transitions, 2006

857 **Donatella Danielli, Nocola Garofalo, and Duy-Minh Nhieu,** Non-doubling Ahlfors measures, perimeter measures, and the characterization of the trace spaces of Sobolev functions in Carnot-Carathéodory spaces, 2006

856 **Vladimir Bolotnikov and Harry Dym,** On boundary interpolation for matrix valued Schur functions, 2006

855 **Yevgenia Kashina, Yorck Sommerhäuser, and Yongchang Zhu,** On higher Frobenius-Schur indicators, 2006

854 **Noam Greenberg,** The role of true finiteness in the admissible recursively enumerable degrees, 2006

853 **Joachim Krieger,** Stability of spherically symmetric wave maps, 2006

852 **Viorel Barbu, Irena Lasiecka, and Roberto Triggiani,** Tangential boundary stabilization of Navier-Stokes equations, 2006

851 **Jie Wu,** On maps from loop suspensions to loop spaces and the shuffle relations on the Cohen groups, 2006

850 **Siegfried Echterhoff, S. Kaliszewski, John Quigg, and Iain Raeburn,** A categorical approach to imprimitivity theorems for C^*-dynamical systems, 2006

849 **Katsuhiko Kuribayashi, Mamoru Mimura, and Tetsu Nishimoto,** Twisted tensor products related to the cohomology of the classifying spaces of loop groups, 2006

848 **Bob Oliver,** Equivalences of classifying spaces completed at the prime two, 2006

847 **Eric T. Sawyer and Richard L. Wheeden,** Hölder continuity of weak solutions to subelliptic equations with rough coefficients, 2006

846 **Victor Beresnevich, Detta Dickinson, and Sanju Velani,** Measure theoretic laws for lim–sup sets, 2006

845 **Ehud Friedgut, Vojtech Rödl, Andrzej Ruciński, and Prasad V. Tetali,** A Sharp threshold for random graphs with a monochromatic triangle in every edge coloring, 2006

844 **Amadeu Delshams, Rafael de la Llave, and Tere M. Seara,** A geometric mechanism for diffusion in Hamiltonian systems overcoming the large gap problem: Heuristics and rigorous verification on a model, 2006

843 **Denis V. Osin,** Relatively hyperbolic groups: Intrinsic geometry, algebraic properties, and algorithmic problems, 2006

842 **David P. Blecher and Vrej Zarikian,** The calculus of one-sided M-ideals and multipliers in operator spaces, 2006

841 **Enrique Artal Bartolo, Pierrette Cassou-Noguès, Ignacio Luengo, and Alejandro Melle Hernández,** Quasi-ordinary power series and their zeta functions, 2005

840 **Sławomir Kołodziej,** The complex Monge-Ampère equation and pluripotential theory, 2005

839 **Mihai Ciucu,** A random tiling model for two dimensional electrostatics, 2005

838 **V. Jurdjevic,** Integrable Hamiltonian systems on complex Lie groups, 2005

837 **Joseph A. Ball and Victor Vinnikov,** Lax-Phillips scattering and conservative linear systems: A Cuntz-algebra multidimensional setting, 2005

836 **H. G. Dales and A. T.-M. Lau,** The second duals of Beurling algbras, 2005

835 **Kiyoshi Igusa,** Higher complex torsion and the framing principle, 2005

834 **Kenichi Ohshika,** Kleinian groups which are limits of geometrically finite groups, 2005

For a complete list of titles in this series, visit the AMS Bookstore at **www.ams.org/bookstore/**.